Debating Darwin

DEBATING DARWIN

Robert J. Richards and Michael Ruse

THE UNIVERSITY OF CHICAGO   CHICAGO AND LONDON

Robert J. Richards is the Morris Fishbein Distinguished Service
Professor in History of Science at the University of Chicago, where he is
professor in the department of History, Philosophy, and Psychology and
in the Committee on Conceptual and Historical Studies of Science and
directs the Fishbein Center for the History of Science and Medicine.
His books include, most recently, *Was Hitler a Darwinian? Disputed
Questions in the History of Evolutionary Theory*, also published by the
University of Chicago Press. **Michael Ruse** is director of the Program
in the History and Philosophy of Science at Florida State University.
His books include *The Gaia Hypothesis: Science on a Pagan Planet*,
also published by the University of Chicago Press.

The University of Chicago Press, Chicago 60637
The University of Chicago Press, Ltd., London
© 2016 by The University of Chicago
All rights reserved. Published 2016.
Printed in the United States of America

25 24 23 22 21 20 19 18 17 16   1 2 3 4 5

ISBN-13: 978-0-226-38442-9 (cloth)
ISBN-13: 978-0-226-38439-9 (e-book)

DOI: 10.7208/chicago/9780226384399.001.0001

Library of Congress Cataloging-in-Publication Data
Names: Richards, Robert J. (Robert John), author. |
Ruse, Michael, author.
Title: Debating Darwin / Robert J. Richards and Michael Ruse.
Description: Chicago : The University of Chicago, 2016. |
Includes bibliographical references and index.
Identifiers: LCCN 2015048965| ISBN 9780226384429
(cloth : alk. paper) | ISBN 9780226384399 (e-book)
Subjects: LCSH: Evolution (Biology) | Darwin, Charles, 1809–1882.
Classification: LCC QH366.2 .R5238 2016 | DDC 576.8/2—dc23
LC record available at http://lccn.loc.gov/2015048965

♾ This paper meets the requirements of
ANSI/NISO Z39.48–1992 (Permanence of Paper).

CONTENTS

PREFACE

The British naturalist Charles Robert Darwin (1809–82), author of *On the Origin of Species by Means of Natural Selection or the Preservation of Favoured Races in the Struggle for Life*, is rightfully known as the "father of evolution." In his lifetime, Darwin's accomplishments were recognized and appreciated. At his death he was buried in that British Valhalla, Westminster Abbey, where he lies today, next to the great Isaac Newton. He is still respected and venerated, both publicly and professionally. In the world of everyday life, his bearded face peers out from the back of the British ten-pound note. In the world of science, he is recognized as one of the truly great thinkers whose achievements are the foundation for much of contemporary biology.

From the very beginning, Darwin and his ideas were highly controversial. During his lifetime the religiously orthodox began an attack that has continued to the present day, especially in the United States. Though some churchmen have made accommodation to evolutionary theory, religious fundamentalists still regard Darwin as the enemy; and they are often abetted by conservative politicians. In the scientific community, no serious biologist doubts what might be called the fact of evolutionary descent, though researchers still debate the precise role of natural selection in producing species change. Among social scientists, humanists, and philosophers, the reaction to Darwinian theory is mixed; few deny its power in explaining the development of plant and animal species, but many would hesitate to apply evolutionary considerations to account for human behavior and social relationships.

Given the magnitude and reaction to Darwin's theory, it is hardly surprising that historians and philosophers of science have taken a deep interest in his intellectual development and the precise nature of his accomplishment. They have been aided in their research by Darwin's own habits of mind—he retained almost every scrap of paper to which he put pen. The collection of manuscripts at Cambridge Library and other archives has allowed scholars to follow Darwin in the production of his ideas; and much of this material is now in print or online. The Cambridge edition of Darwin's correspondence, for example, has now reached volume 22, with at least another ten planned; and many of his manuscripts have been digitized and made available on the Internet.

You might expect that with all the resources now available to Darwin scholars a consensus would have been reached about the nature of his achievement. Certainly there is agreement about the broad outlines. We know, for example, about when and under what conditions Darwin came to endorse the transmutation of species, and what stimulated him to formulate the principle of natural selection. We can track with some assurance the fate of his religious convictions, and be confident about his intention to bring human beings under the explanatory framework of his theory. But the facts of Darwin's development and the claims of his theory do not speak for themselves. Or rather, they speak for themselves only when the historian has put them in proper context and the philosopher has entered into the mind of Darwin to understand how he conceived these facts and claims. With respect to the interpretative framework and the conclusions to be drawn about Darwin's intentions, we, the authors of this book, do differ and passionately so. In the pages that follow, our differences will be on vibrant display: our arguments will be pointed and the responses aggressive. Our dispute has been of long standing, but it has not tainted our friendship.

It might be thought that our differences are essentially a function of disciplinary boundaries. One of us, Ruse, has always been located in departments of philosophy. The other, Richards, has long been a member of departments or centers of history. Hence, it might be supposed that the disagreements come from talking past each other, as the philosopher wants to stress unadorned, timeless concepts and the historian wants to place everything in time-bound culture. This is not so. We both take on questions of historical context and philosophical interpretation, and recognize that our disagreements are more profound and more interesting than simple disputes about disciplinary methods. We are not talking past each other but right at each other. Yet each comes to quite different conclusions and thinks the other has simply been wandering in the intellectual wilderness.

Darwin was British-born and educated in the English system. Apart from a five-year voyage on HMS *Beagle* that took him around the globe, he spent the whole of his life in Britain. Is this the essential key to the man and his work? One of the authors, Ruse, thinks that it is, absolutely and completely. He sees Darwin's science as British as (let us say) Lord Palmerston's foreign policy or Charles Dickens's fiction or Joseph Paxton's Crystal Palace built for the Great Exhibition. The other author, Richards, argues that it is not Darwin's physical geography that essentially matters, but his mental geography, which extends far beyond British shores. It was, after all, the German Romantic Alexander von Humboldt's account of his travels to the new world that led Darwin to embark on his own romantic adventure. Richards believes that to ignore the impact of the German Romantics and their legacy—especially that legacy transported to England and traveling under the guise of British names—would be to miss the significance of Darwin's achievement in the *Origin of Species* and the *Descent of Man*.

This is our disagreement. Was Charles Darwin quintessentially British, or was his attitude thoroughly cosmopolitan, encompassing as well ideas from German Romantic sources? More specifically, this is a debate about such topics as mechanism or mind in nature; teleology faux or real; human beings deluded about their moral character or intrinsically moral. And what does this tell us about the present? We are both sufficiently indoctrinated into modern historiographical practices that we rear with horror at the thought of writing something that simply tells a story of progress from the mistaken past to the enlightened present; but we are both evolutionists, and we think that, in culture as in biology, in order to understand the present you must understand the past. Hence we do not look upon this clash as an exercise in self-indulgence, two good friends simply having a vigorous game of intellectual handball.

We think that what we have to say matters and that, depending on the side you think is the more convincing, so will you view evolutionary thought and its implications today. We will be especially keen to indicate how these historical matters impinge on our understanding not only of nature writ large, but also on human nature and especially on the moral character of our species. The conflagrational disputes over sociobiology, evolutionary psychology, and selfish genes have concerned the way Darwinian theory has construed human nature—indeed, we might ask, can we even speak of a distinctively human nature in the wake of evolutionary considerations? We believe that these disputes will achieve greater clarity when we return to their original site in the work of Charles Darwin.

We had thought that we might be able to write a neutral historical introduction laying out some of the established facts about Darwin and his work. Very quickly we found that this was impossible. In an almost Kantian fashion, as soon as we started to look at the real world, interpretation kept rushing in. So we have set about telling the story in our own ways, although we have constantly exchanged ideas and drafts in order to focus our own thinking and to sharpen our points of disagreement; we do, though, provide a shared timeline of the main events. After each of our essays, we make a concise response to the other's arguments. In the epilogue to this book, we join together to trace the consequences of Darwin's accomplishment for the development of evolutionary theory in the period of the late nineteenth century to the present. We are especially attentive to what Darwinian theory implies for that most characteristic of human traits, conscious thought and religious aspiration. This essay thus seeks to discover what is still living and vital in the ideas that have given rise to modern biology—yet more, the role of those ideas in coming to understand ourselves.

We are indebted to David Sepkoski, Mark Borrello, and Gregory Radick, who patiently read earlier drafts of our contributions. They kept their criticisms jagged and merciless. Despite their rough treatment, we are deeply grateful for the application of their incisive considerations. We wish to thank, as well, our editor at the University of Chicago Press, Karen Darling. Her own reading helped each of us focus efforts to greater effect. She did not take sides and kept her amusement impartial. The Press's referees made decisive suggestions, for which our thanks is due. We are grateful to the John Templeton Foundation for its financial support in our enterprise. Ruse would also like to thank the Stellenbosch Institute for Advanced Study in South Africa—noting especially its director Hendrik Geyer and his staff—that provided a home while he completed his share of this exchange. Finally, readers will find that understanding our arguments and judging our differences demand a very close reading of Darwin's texts, both those published and those unpublished (in his lifetime). We strongly recommend that the reader make full use of two extremely helpful websites: John van Wyhe's *Darwin Online*— http://darwin-online.org.uk/; and David Kohn's *Darwin Manuscripts Project*—http://www.amnh.org/our-research/darwin-manuscripts -project. Not only will the reader find at hand all of Darwin's published writings, in their many editions, but also the vital, unpublished sources.

TIMELINE

1688   The Glorious Revolution—the Catholic King James II is deposed and the Protestant monarchs William III and Mary II assume the throne

1712   Invention of the Newcomen engine, used to pump water out of mines

1749   Birth of Johann Wolfgang von Goethe

1757   David Hume's *Natural History of Religion* is published

1759   Josiah Wedgwood founds pottery works

1760   Robert Bakewell takes over family farm in Leicestershire and starts program of intensive breeding

1776   Adam Smith's *Wealth of Nations* is published

1769   Birth of Alexander von Humboldt

1761   Opening of Bridgewater Canal, taking coal to Manchester

1781   Immanuel Kant's *Critique of Pure Reason* is published

1785   Invention of first power loom, by Edmund Cartwright

1789   Start of the French Revolution

1790   Immanuel Kant's *Critique of the Power of Judgment* is published

1790   Johann Wolfgang von Goethe's *Metamorphosis of Plants* is published

1794   Erasmus Darwin publishes the evolutionary work *Zoonomia*

1798   Thomas Robert Malthus publishes *An Essay on the Principle of Population*

1799–1804 Alexander von Humboldt and Aimé Bonpland travel to the Americas

1802   William Paley publishes *Natural Theology*

1804   Napoleon crowns himself Emperor of France

1807   Slave trading made illegal in the British Empire

1808   Johann Wolfgang von Goethe's *Faust*, part 1, appears

1809   Jean-Baptiste Lamarck's *Philosophie Zoologique* is published

1809   Birth of Charles Robert Darwin

1813   Robert Jameson publishes Georges Cuvier's *Essay on the Theory of the Earth* (translation of 1812 French edition)

1813   Napoleon defeated by the allies at the Battle of Leipzig, where 600,000 soldiers clashed

1815   Battle of Waterloo, Napoleon finally defeated

1817   Georges Cuvier publishes *Le Règne Animal* (*The Animal Kingdom*) stressing conditions of existence

1817–24   Johann Wolfgang von Goethe's collection *Zur Morphologie* is published

1818   Darwin enrolls as boarder at Shrewsbury School

1818–29   Translation of Alexander von Humboldt's seven-volume *Personal Narrative of Travels to the Equinoctial Regions of the New Continent, 1799–1804* is published

1825   Darwin begins medical studies at Edinburgh University

1828   Darwin enrolls at the University of Cambridge to start BA with the intention of becoming an Anglican clergyman

1828   Carl Gustav Carus's *Von der Ur-Theilen des Knochen- und Schalengerüstes* is published

1830   Opening of first steam passenger railway, between Manchester and Liverpool

1830   John F. W. Herschel publishes *A Preliminary Discourse on the Study of Natural Philosophy*

1830–33   Charles Lyell publishes the *Principles of Geology*

1831   Darwin joins HMS *Beagle* under the captaincy of Robert Fitzroy

1831	First meeting (in York) of the British Association for the Advancement of Science
1832	First Reform Act (Darwin's uncle Josh, father of Emma Wedgwood, becomes a member of the new parliament)
1833	Abolition of slavery throughout the British Empire
1834	New Poor Law, creating "unions" of workhouses, so unpleasant that the poor would do anything to avoid them
1835	The *Beagle* visits the Galapagos Archipelago in the Pacific Ocean
1836	The *Beagle* returns to England
1837	In the spring, influenced by the British Museum ornithologist John Gould's identification of three types of Galapagos mockingbird as good species, Darwin becomes an evolutionist
1837	William Whewell publishes *The History of the Inductive Sciences*
1838	At the end of September, Darwin reads Malthus and discovers natural selection
1839	Early in the year, Darwin marries his first cousin Emma Wedgwood
1839	Darwin's *Journal of Researches of the Voyage of the Beagle* is published
1840	Rowland Hill starts the penny post
1840	William Whewell publishes *The Philosophy of the Inductive Sciences*
1842	Darwin writes out the first "Sketch" of his theory, some 35 pages
1844	Robert Chambers publishes *The Vestiges of the Natural History of Creation* anonymously
1844	Darwin expands his 1842 sketch into a 230-page manuscript, the "Essay"
1846	Darwin begins his study of barnacles
1849	Richard Owen's *On the Nature of Limbs* is published
1851	The Great Exhibition, celebrating Britain's supremacy in industry and technology

1852	Herbert Spencer starts writing on evolutionary topics
1854	Darwin finishes with his four volumes on extant and extinct barnacles, and turns back to evolutionary topics
1856–58	Darwin works on manuscript to be called *Natural Selection*, which in abbreviated form becomes the first part of the *Origin of Species*
1858	Alfred Russel Wallace sends to Darwin his paper on evolution
1859	Toward the end of the year, Darwin publishes the *Origin of Species*
1863	Thomas Henry Huxley publishes *Man's Place in Nature*
1871	Darwin publishes the *Descent of Man*
1882	Darwin dies and is buried in Westminster Abbey
1900	Mendel's thinking on heredity is rediscovered
1930	Ronald A. Fisher publishes *The Genetical Theory of Natural Selection*
1931	Sewall Wright publishes "Evolution in Mendelian Populations"
1964	William D. Hamilton publishes "The Genetical Evolution of Social Behaviour"
1975	Edward O. Wilson publishes *Sociobiology: The New Synthesis*
1976	Richard Dawkins publishes *The Selfish Gene*
1995	Daniel Dennett publishes *Darwin's Dangerous Idea*
2012	Thomas Nagel publishes *Mind and Cosmos: Why the Materialist Neo-Darwinian Conception of Nature Is Almost Certainly False*
2015	Jerry Coyne publishes *Faith vs. Fact: Why Science and Religion Are Incompatible*

CHARLES DARWIN: GREAT BRITON

Charles Darwin was first and foremost a scientist, a very great scientist, who not only made scientifically plausible the idea of organic evolutionary change but who came up with natural selection, what today's professional scientists generally consider to be the chief motive force of such change. Yet from the first, as Darwin himself recognized, his thinking was always more than just about scientific explanations of the organisms occupying the physical world. His thinking pointed the way to a new or revived philosophical perspective on reality. A harsher, less-comfortable one than that he inherited. The popular-science writer and ardent atheist Richard Dawkins has written:

> In a universe of blind physical forces and genetic replication, some people are going to get hurt, other people are going to get lucky, and you won't find any rhyme or reason in it, nor any justice. The universe we observe has precisely the properties we should expect if there is, at bottom, no design, no purpose, no evil and no good, nothing but blind, pitiless indifference. As that unhappy poet A. E. Houseman put it:

> > For Nature, heartless, witless Nature
> > Will neither know nor care.

> DNA neither knows nor cares. DNA just is. And we dance to its music.[1]

As a staid and very respectable Victorian, Charles Darwin would have been horrified at the frenzied polemics that characterize the writings of the so-called New Atheists. Whatever his personal beliefs, he would never have flaunted his thinking in such a crude and public way. It is doubtful also whether Darwin ever reached quite the state of naturalistic nihilism expressed by Dawkins. Even if he took us all of the way, it is certainly not my claim that Darwin unaided took us to this new world. Internal issues in religion like so-called higher criticism (looking at the Bible as a human-written document) played a crucial role, as did social factors like the move from the land to the city demanding new ideologies for new types of existence. But Darwin's work pointed that way, and he knew it and pursued it. If like Moses and the Promised Land he never quite arrived, he beat the path toward it, consciously and intentionally. Darwin changed not just science; he changed philosophy also, and this is the world in which we now live.

Such is my claim in this, my section of this book. Moreover I argue that Darwin did all of this within a tradition on which he drew. A tradition that in many respects was quintessentially English, the land of his birth, but that was more broadly British, not only because Darwin was in part educated north of the border, but because Darwin always drew heavily on thinking that came from the so-called Scottish Enlightenment. In short, I argue that although Darwin was a great revolutionary—and I bow to no one in my belief that he made major advances in our understanding of the empirical world—he was not a rebel. He did not repudiate his past, hating and trying to destroy and eliminate that from whence he came. It was rather that he took what was offered and then rearranged and transformed the elements into an altogether new picture. Darwin's work was like a kaleidoscope. The pieces were there. Darwin shook them up and made something different. But where did the pieces come from that I claim were so important in Darwin's past? I argue—and here I would stress that I am being totally unoriginal and simply drawing on what one finds in any good textbook—that the Britain into which Darwin was born at the beginning of the nineteenth century had two major elements or themes or traditions. It was his good fortune to be able to draw on both elements and his genius to do with them what he did.

The one element is what we might with reason call the conservative element, the Tory side to Britain. This is the world of the king (George III and the Prince Regent, the future George IV) and of his supporters, political, military (including naval), and most of all clerical. It is the world of landowners, but usually not the biggest men. They were more the leaders in the villages that one finds in the novels of Anthony Trollope (although he was writing a little later), men like Wilfred Thorne, the squire of St. Ewold's in *Barchester Towers*. It is the world of the Church of England parson, the world (again in *Barchester Towers*) of Archdeacon Theophilus Grantly. And it is very much the world of England's two ancient universities, Oxford and Cambridge. The clerical world and the academic world were truly but one, for to graduate from the universities one had to be a paid-up, believing member of the Church of England and most of the teachers, the "dons," at Oxford and Cambridge had taken holy orders. To refer one more time to Trollope's great novel, remember that the man who becomes Dean of Barchester, Francis Arabin, is a fellow of Lazarus College (a thinly veiled portrait of Christchurch) and a sometime professor of poetry at Oxford.

The other element is what we might with equally good reason call the liberal element, the Whig side (after the Reform Bill of 1832 joined by the Radical side) to Britain. Their leaders were the great landowners, men like the Duke of Omnium in Trollope's political novels. Somewhat paradoxically, they were often joined by the bishops of the Church of England. Bishoprics are bestowed by the government of the day, sometime Whig or liberal. The politicians wanted supporters not the thanks of the village priests. The plot of *Barchester Towers* revolves around the fact that Archdeacon Grantly, firmly Tory, does not get to follow his father into the see of Barchester. The post goes instead to the Whig Bishop Proudie. The leaders of the Whigs were allied with the men of industry. Whereas the Tories inclined toward protectionism, looking to the interests of the rural leaders—the notorious Corn Laws enacted after the Napoleonic Wars were the epitome of such inclinations, designed as they were to keep high the value of homegrown grains—the Whigs inclined toward free trade, something that opened up markets for the products, initially and overwhelmingly cotton but later moving more toward manufactured goods in iron and nonferrous metals, flowing from the labors of those directed by the leaders of industry. There was often no conflict between the interests of the big landowners and the industrialists, because the former owned valuable coal and mineral deposits on which the ever-increasing number of factories very much depended.

I shall argue that both of these elements had beliefs and ideologies, secular and sacred, that spoke to their interests. I shall argue also that Darwin almost uniquely was in a position to draw on both sides and that he did. Darwin's genius may be a mystery—why should a young man of somewhat modest gifts (in areas like linguistic or mathematical abilities), who was born to a life of ease, end by doing so much? The influences from the culture in which he was reared, the sources on which he drew, are no mystery. They span the spectrum of ideas and beliefs that formed and molded the society into which he was born. And it was because of this that Darwin was set on his life's quest, one that transformed the life sciences and—as encapsulated in the quoted passage by Richard Dawkins—took us to the world of today, a world that many still resist but that in the end closes off the world of yesterday, the world into which Charles Darwin was born.

BRITAIN BEFORE DARWIN

The "Glorious Revolution" of 1688 saw the dethroning of the Catholic king, James II, and the accessions of his Protestant daughter and son-in-law, Mary II and William III. As importantly, it saw the real beginnings in Britain of "constitutional monarchy," where increasingly parliament had an effective voice in the running of the country. When James's Protestant daughters, Mary and then Anne, failed to produce heirs, the throne was handed over to the rather dull, but safely Protestant, German royal family from Hannover, whose dynasty lasted through the life of Charles Darwin, ending only with the death of Queen Victoria in 1901. Uninspiring though the family may have been, it ruled over a country that went at the beginning of the eighteenth century from the fringe of Europe to ending the nineteenth century as the greatest power that the world had ever seen, with a quarter of the globe colored red for the British Empire. No one single causal factor can be isolated for this growth, but a major factor was the freeing of the country from the autocratic power of monarchs whose chief interests would have been in preserving the structure of the society that had promoted them to the pinnacle. With others now having not just an interest in the fortunes of the country, but with real power and possibilities of molding things to their own ends, almost uniquely the country had reasons to promote stability and the chance to move forward in new directions. Combine this with massive increases in scientific knowledge in the seventeenth century, often geared to practical ends, and the unrivaled natural gifts of the land—ready supplies of fuel, an abundance of needed minerals, rivers and seas for easy transport, a temperate climate, and much more—and Britain was able to seize the chance and build that industrial land on which its future fortunes were to be based. Ours is a story about one part of that great and progressive change.[2]

From Farm to Factory

If the metaphor of the Scientific Revolution is the timepiece—in the words of Robert Boyle, the world is "like a rare clock, such as may be that at Strasbourg, where all things are so skillfully contrived that the engine being once set a-moving, all things proceed according to the artificer's first design"[3]—then the metaphor of the British Industrial Revolution of the eighteenth century is the Newcomen engine[4] (see figure 4). Making its first appearance in the second decade of the century, it transformed mining as it worked its steady pace to suck the water out of the tunnels far below and made possible ever-greater exploitation of the minerals and fuels there for the asking. One may question whether, as has been suggested, its feedback processes—heated steam expanding and then bringing on squirts of cold water and consequent condensation and contraction—are mirrored by the economy of the day—laissez-faire leading to overproduction, contraction, and ever-newer opportunities, all driving the country forward—whereas the never-deviating, endless motions of the clock mirrored the fixed and stifling rules of countries beneath the yokes of all-powerful monarchs.[5] What is beyond question is that the engine and the many subsequent inventions—especially those that transformed the production of cotton—lay at the heart of the great changes that ran through almost every part of the British Isles.

Yet to focus first on industry is to get ahead of ourselves. Napoleon Bonaparte said that "an army marches on its stomach." The same can be said of countries, so let us start there. Britain, England particularly, saw major changes in agriculture and food production in the eighteenth and early part of the nineteenth century. The amount produced increased hugely and at the same time the labor required stayed constant, to the extent even of freeing for other opportunities numbers who hitherto had had some connection with the land. There were several reasons for this, although whether cause or effect is often hard to discern. New crops were being introduced, notably clover and turnips. The latter particularly played a crucial role in enabling farmers to feed their livestock over winter without the need for annual mass slaughter at the end of the summer. Methods of livestock improvement were being discovered and refined. Above all it was realized that selective breeding was the key to success. With these changes, the social structure of rural Britain was being changed. To this point, people working on the land had followed rules and practices that reached back into medieval times, with small-holders tilling strips of land that rotated crops, with common land for grazing, and with woodlands for wood collecting and foraging. Now, land was being "enclosed," cut off from public ownership and made the property of individuals, and marginal members of society, who had before subsisted on traditional rights of gleaning and keeping a cow or two on common land and finding fuel in the woods, were either reduced to the roles of employed day laborers or encouraged to leave and move to the ever-growing towns and cities.

Increasingly work was becoming available in the urban centers, particularly the new towns and cities of the British Midlands and the North—Birmingham, Manchester, Liverpool, Leeds, Newcastle, and up into Scotland. Obviously this more industry-focused labor was not something that appeared overnight, but slowly and surely implements and machines were introduced at various stages of the process and as slowly but surely it became more and more efficient to collect workers all in one place and to impose on them the rules and restrictions of the modern workplace. The reasons for change and where and how it occurred were manifold and often complex, but one thing does stand out, namely, that increasingly fossil fuel was used to supplement or replace hand labor. In a word, coal. Its availability in Britain was perhaps the major factor in the move to industrialism and the amount mined grew almost exponentially in the century and a half beginning in 1700. The amount mined fueled the changes but at the same time demanded changes, especially in devising ever-more-efficient pumps to remove water from the ever-deeper shafts being dug. And there was a ripple effect. Carrying something like coal is far, far easier by water than by land, and so there was an improving of already-existing waterways and the digging of a network of new canals all over the country. Within a year (1761), a new canal (the Bridgewater) linking Manchester with a colliery a few miles outside the city dropped the price of coal by half.

The changes led to new patterns of everyday life and most particularly to an explosive growth in the population. Down on the farm, the younger generation basically had to wait until the older generation could no longer do the daily work. There was therefore strong incentive to postpone marriage and a family until one could take over and build a life for oneself. In the town or city, working in a factory, the highest wage period came early, and so there was much less reason for restraint. Essentially this meant that the childbearing time was longer and so families grew in size. The biology was reinforced by culture, because a lot of the new industries put a premium on the work of women and children, and thus a larger family equaled a more prosperous family.

Making Sense of Change

Naturally these changes attracted the attention of the theorists, and it is in this time that we see the birth of the science of political economy. Even today, the Scot Adam Smith (figure 2) commands respect. He was the theoretician of the factory and its functioning, introducing one of the all-time, best-known, and most powerful metaphors: "the division of labor," or "labour" as he spelled it.[6] Taking the example of the manufacture of pins, Smith argued that a man working on his own, doing everything, would make but a few dozen, if that, a day. But divided into a team, with each doing his allotted task—grinding, polishing, and so forth—literally thousands a day can be produced. There is no magic to this. It is more efficient that each person perfect his or her own skill and do it time in and time out, passing on the semi-finished product to the next down the line until the whole job is finished. Smith was also keen on transport, especially by water. "Six or eight men . . . by the help of water-carriage, can carry and bring back in the same time the same quantity of goods between London and Edinburgh, as fifty broad-wheeled waggons, attended by a hundred men, and drawn by four hundred horses."[7]

And above all, introducing yet another of the famous metaphors of British culture, Smith lauded the virtues of self-interest, where everyone is seen in the rather unkind words of the author of *Peter Pan*, playwright J. M. Barrie, as a "Scotsman on the make." We naturally tend to promote the industry of the land within which we live, for the obvious reason that we will be better off and more secure in a prosperous nation rather than otherwise. By so doing, an individual "generally, indeed, neither intends to promote the public interest, nor knows how much he is promoting it. By preferring the support of domestic to that of foreign industry, he intends only his own security; and by directing that industry in such a manner as its produce may be of the greatest value, he intends only his own gain, and he is in this, as in many other eases, led by an invisible hand to promote an end which was no part of his intention."[8] Not that this end will necessarily be only of benefit to the individual. "By pursuing his own interest he frequently promotes that of the society more effectually than when he really intends to promote it." Concluding sardonically: "I have never known much good done by those who affected to trade for the public good." The ultimate power, the deity—"The invisible hand"!—has seen to it that individual self-regard spells benefits for all. As Smith put it somewhat more pithily: "It is not from the benevolence of the butcher, the brewer, or the baker that we expect our dinner, but from their regard to their own interest."[9]

Then at the end of the eighteenth century came Parson Malthus (figure 3). Appalled at the naive optimism that he saw emanating from the continent—and no doubt frightened near to death by the dreadful consequences (the "Reign of Terror") to which he thought it had led—and certainly mindful of the incredible population explosion now in full flight in Britain, Thomas Robert Malthus (he was generally known as "Bob") published a pamphlet that in succeeding editions grew into a full-sized book, in which he drew attention to the dire expectations that we should expect from unrestrained sexual activity and the production of ever-more mouths to feed.[10] Food can be produced and increased only according to an arithmetic scale: 1, 2, 3, 4 . . . Population numbers however have the potential to go up at a geometric rate: 1, 2, 4, 8 . . . This can lead only to strife and conflict. Eventually there will be fights among humankind for territory and food. Introducing yet another of the famous metaphors that will control future discussion, Malthus suggested that the young of a tribe will be expelled and go searching for their own space and provisions. "And when they fell in with any tribes like their own, the contest was a struggle for existence, and they fought with a desperate courage, inspired by the rejection that death was the punishment of defeat and life the prize of victory."[11]

The Levers of Power

What kind of country was Britain in the eighteenth and early nineteenth century? Since the Act of Union of 1707, England (and Wales) and Scotland were one country, although with differences especially in law. (The two countries had shared the crown since 1603.) The Glorious Revolution pushed the country toward democracy, and this was crucially important. But it must be allowed that it was hardly a democracy in a sense that we would understand or appreciate. The powerful aristocrats (remember, usually members of the Whig party) balanced the throne and its supporters—the earlier-mentioned smaller landowners (squires) and military and clergy—and these noblemen held much power both in the House of Lords (where they sat by hereditary right) and the House of Commons (where they sent members who were chosen by them and beholden to them). This meant that such political power was in the hands of men who were usually rich because of owning many acres, and that (especially since the other side tended to be even less sympathetic) the men of industry and commerce were too often excluded from the great decisions of state. New growing towns and cities like Birmingham had no members of parliament, whereas some rural ridings with very few inhabitants ("rotten boroughs") returned members chosen by the aristocratic patrons. Of course in reality there was much more movement of power and interest up and down the classes—mention was made earlier of the fact that the interests of the landowners were often at one with the interests of the industrialists—but it was not until 1832 that the first of the Reform Acts was enacted, starting the real redistribution of power among the classes of the country.

Interestingly from our perspective this did not necessarily spell improvement as we might judge it. With the move to cities, with men and women plying for work in a personally indifferent market, with growing distances between masters and employees—no longer did one have the traditional squire-yokel relationship—those newly empowered were keen to keep the poor rates to a minimum. So great new workhouses were erected, intended to keep together soul and body, but to be so unpleasant that the indigent would do all in their powers to avoid falling for mercy on the state. Shades of *Oliver Twist* (first published in book form in 1838)! Although many, especially the newly empowered industrialists, would have derided such sentimentality and argued that the Malthusian facts speak for themselves. The population of England doubled between 1781 (about 7 million) and 1831 (about 14 million). Glasgow grew from 62,000 in 1791 to 202,000 in 1831; Manchester from 30,000 pre-1800 to 182,000 in 1831; and Birmingham from 42,000 in 1778 to 144,000 in 1831. London from a million and a quarter in 1801 to over 2 million in 1831.[12] Industrialists were often torn over facts such as these. On the one hand, they necessitated heavy payments to support the indigent. On the other hand, they offered ready supplies of very cheap labor. It is noteworthy how one and the same person could be dreadfully upset by the export of African slaves to the New World, and yet indifferent to the needs of the poor of his own country. Josiah Wedgwood founded the great pottery works, and he and his family were leaders in the fight against the slave trade. Famously he produced a medallion of a chained slave imploring, "Am I not a man and a brother." As famously, speaking of his own workers, his avowed aim was "to make such machines of the men as cannot err." Of their taking time off to go to fairs, he threatened, "I would have thrashed them right heartily if I could."[13]

What of the role of religion in all of this? On coming to the throne in 1558, Elizabeth—"Good Queen Bess"—had fixed Britain as a Protestant country, and for all of the troubles with proselytizing Jesuits and rambunctious Puritans, not to mention the horrors of the civil war in the middle of the seventeenth century and the appearance of dissenters like the Baptists and the Quakers, by the beginning of the eighteenth century the Established Church of England, the Anglican Church, was firmly in control. It was said that the Church of England was (government-appointed Whig bishops aside) the Tory party at prayer. There was good reason for this, for time and again the local incumbent was a brother or younger son or other close relative of the local landowner, and being given a (lifetime) tenancy of a parish was considered both socially and financially an appropriate role for a gentleman. Theologically the Anglican Church pointed to a comfortable conservative perspective—one that would have disdained a wild lurch to the right as much as it would have deplored the left-wing movements of the twentieth century. The "Elizabethan Settlement" or "Compromise," steering between the authority claimed by the Catholic hierarchy and sola scriptura of the Calvinists, put a heavy emphasis on traditional forms, on stability, on the paternalistic obligations of those in authority. Often it was leading Tory laypeople who were the leaders in shortening the workday and preventing use of women and children for the worst kind of labor.

Not for the Anglicans were the speculative flights of continental thinkers—Schleiermacher and a feeling of "absolute dependence" and that sort nonsense—but a comfortable empirical approach to the mysteries of creation, as expounded above all in the turn-of-the-nineteenth-century textbooks of Archdeacon William Paley of Carlisle (figure 1). With respect to revealed theology, in *A View of the Evidences of Christianity*, Paley assured us that the willingness of the disciples to die for their faith confirms the authenticity of the Gospel miracles and hence the divinity of Christ.[14] With respect to natural theology—an Anglican favorite since the sixteenth century—there was to be no intellectual chicanery with flashes of unsound Cartesian brilliance like the ontological argument.[15] It is all a matter of design; although it is revealing how Paley in his *Natural Theology* showed his own old-fashioned roots by seizing on a watch as the paradigmatic example of intelligence at work.[16] No matter. So complex and functioning an entity cannot have been formed by chance. Likewise does the eye bear testimony to a designer as much as does a telescope. There has to be a good all-powerful God. One who has ordered society as it is and with which we should not mess.

Providence versus Progress

There were winds of change. Science, for all of Galileo's troubles, was almost always done by sincere believers, but increasingly it made improbable many of the more outlandish claims about the supernatural. And foreign travel, especially to the East, opened many an eye wider than hitherto thought possible. Inhabitants of these lands were not all savages, they had sophisticated religions of their own, and not a murmur could be found of the doings of Jesus of Nazareth! Could it be that Christianity was not true? There were two responses to this question.[17] One to draw the line. The rot had gone far enough and must be stopped. The heart must rule the head. In England, and then of course increasingly across the Atlantic, we have the Methodists. The message was simple: "Believe and ye shall be saved." For all that John Wesley, the leader, was an educated man, he did not find salvation while exploring the friendly fields of English natural theology, nor did he walk through these with the thousands who flocked to hear his message. His heart was "curiously warmed," and the same was true for the many that followed him. The other response was to follow on down the path of reason and empirical evidence. To let the head have full sway. This did not, at least this did not in Britain, lead at once to ardent atheism. There was no proto Richard Dawkins. But on both revealed and natural theological grounds people did start to have doubts, and there was a move to what is known as "deism." This is the idea of God as an unmoved mover, who set the world in motion and now sits back and watches his handiwork. It is distinguished from "theism," generally a term restricted to the Abrahamic religions (Judaism, Christianity, and Islam) where God is immanent and willing and able to intervene in His creation. A world, that is, of miracles.

How can one best characterize the two world pictures? Of course, there were all sorts of debates about reading the Bible, but the fights that we see today in America over literalism were not really the focus of difference. These fights are very much the end points of theological inventions in the New World in the nineteenth century. It is more profitable to cast matters in eschatological terms. For believers, the key notion is that of a Providential God. This is a God who will guarantee salvation and eternal life if only one believes and lets one's sins be washed away by the Blood of the Lamb. This is an evangelical religion, and although it is very much a characteristic of Protestant non-Anglicans (nonconformists or dissenters), it spread up and captured many members of the established church—not all of whom were quite as vile and unctuous as Trollope's Obadiah Slope (the personal chaplain of Bishop Proudie in *Barchester Towers*). The Anglican evangelicals were among the leaders of the move against slavery. So it was not a theology of nonaction, but of recognition that standing alone one was doomed to failure.

Opposing Providence was Progress.[18] This is the belief that one can make the Kingdom of Heaven (literal or metaphorical) here on Earth, through one's own unsupported reason and good will and efforts. Education, technology, medicine, agriculture, politics—all can be made better by men and women using their powers properly. Note that it is just as human-focused as is Providence. It is rather that the means to glory are very different. It is well know that, in France, Progress became the philosophy of the day, particularly among the so-called philosophes. It was against one of the leaders of the movement, the Marquis de Condorcet, that Malthus first penned his gloomy reflections on population. But for all of the doubters like Malthus—significantly an ordained member of a Christian church— there were many who saw and reflected on the great strides made in eighteenth-century Britain and who were convinced that this was no contingent phenomenon but a pointer to the possibilities and actualities of genuine, lasting improvement for all. In fact, even Malthus himself was not entirely against something akin to Progress. His discussion was framed within a natural theological context, where he saw the struggle as God's way of getting us to take the initiative and try to better ourselves.[19]

And So to Evolution

It is at this point that our story starts to turn toward evolution, the natural development of plants and animals from forms very different and much simpler, perhaps originally from just a few forms that may themselves have developed from inorganic materials by natural—that is, law-bound—causes. In Britain, the first genuine, full-blown evolutionist was the physician Erasmus Darwin (figure 8), grandfather of the hero of our tale, Charles Darwin. This first Darwin, educated in Edinburgh, was no mere country doctor. From the British Midlands, he was at the heart of the new industrialism, friend of some of the greatest movers, himself an inventor and minor scientist, and ardent member of the Lunar Society, a group who gathered once a month in Birmingham to discuss ideas and plans of mutual concern and interest. He was also a poet, much given to expressing his ideas in (what we today rather judge as) not exactly stellar verse. Be this as it may, it is here that we find some of his most elaborate evolutionary effusions.

Imperious man, who rules the bestial crowd,
Of language, reason, and reflection proud,
With brow erect who scorns this earthy sod,
And styles himself the image of his God;
Arose from rudiments of form and sense,
An embryon point, or microscopic ens![20]

Erasmus Darwin's speculations were not based on empirical evidence. He had had some experience of fossils when tunnels were being bored for canals, but overall his knowledge of facts pertinent or otherwise was minimal, to give a generous assessment. From where then came his enthusiasm for evolution? The fact that we humans are so firmly the end point gives the clue. For Erasmus Darwin, the idea of Progress in the sociocultural world translated itself as evolution in the organic world. Darwin was fanatical about Progress. A good friend of Benjamin Franklin, he was an ardent supporter of the Americans in their break with the home country and, until things started to go dreadfully wrong, was no less enthusiastic about the French revolutionaries. Expectedly as a member of the Lunar Society, he was all in favor of technological change, in verse celebrating the triumphs of his fellow men of business and industry:

So with strong arm immortal BRINDLEY leads
His long canals, and parts the velvet meade.[21]

Explicitly and categorically he drew a parallel between the upward path of culture and that of biology, the two notions really being but one. The idea of organic progressive evolution "is analogous to the improving excellence observable in every part of the creation; . . . such as the progressive increase of the wisdom and happiness of its inhabitants."[22]

The Fall and Rise Again of Evolution

The point to be made is that for Erasmus Darwin, the idea of organic evolution was an epiphenomenon on the culture—the British cul-

ture—of his day. And this point is a guide to the history of evolutionary thinking in Britain for the next fifty years, until the middle of
the nineteenth century. With the collapse of the French Revolution
into terror and anarchy, with the subsequent rise of Napoleon and
his territorial ambitions, contained only after twenty years of hard
slog and battle, the Industrial Revolution paused and reactionary
philosophies prevailed. Those with evolutionary yearnings tended to
keep them under cover, especially since there were growing reports
of evolutionary enthusiasm in those all-too-dangerous lands across
the English Channel.

But with the end of the wars, the Industrial Revolution started
again to pick up steam—literally in the second quarter of the nineteenth century as the country was criss-crossed by railways—and
with it came renewed thoughts and hopes of Progress. All of this
burst forth in 1844 with the publication of the anonymously authored *Vestiges of the Natural History of Creation*: a blooming flower
greeted, with a frisson of illicit delight, by many of the general public;
a suppurating sore that horrified the responding scientific community.[23] Known now to have been written by the Edinburgh publisher
Robert Chambers, it was a powerful potpourri of fact and fiction,
drawing on speculations about the origins of the universe (the so-
called nebular hypothesis), wild claims about the appearance of life
on electric batteries (undoubtedly a function of the dirty hands of
the experimenters), happy reflections on the organic-like nature of
patterns ("frost ferns") on windows in winter, selective gatherings
of information about the growing fossil record, careful pickings of
discoveries about embryological development coming from Germany, and the whole underpinned by an almost-casual deism. What
was not casual was the main point. A self-made man, who had
brought steam-driven printing machines to the very successful cause
of mass publication, Chambers had a message to spread. In the same
mode as Erasmus Darwin, evolution was the consequence of, almost
the predetermined necessary corollary of, the cultural notion of Progress. As, thanks to our efforts, things are bound to improve in our
daily lives, so things change for the better in the biological world.

A progression resembling development may be traced in human nature, both in the individual and in large groups of men. . . . Now all
of this is in conformity with what we have seen of the progress of
organic creation. It seems but the minute hand of a watch, of which
the hour hand is the transition from species to species. Knowing what
we do of that latter transition, the possibility of a decided and general
retrogression of the highest species towards a meaner type is scarce
admissible, but a forward movement seems anything but unlikely.[24]

Had he but known it, Robert Chambers was outdated even as he
published, because privately Charles Darwin was already well into
his speculations about evolutionary origins. So let us pause and go to
Darwin himself. The claim now is not that the work of Erasmus Darwin and Robert Chambers was particularly good. By the standards
even of their own days, it wasn't. The claim rather is that eighteenth-
and early nineteenth-century Britain had a rich culture and that evolutionary speculations arose from it rather than despite it. Is this the
pattern we are to see repeated in the sections that follow?

A CHILD OF HIS CLASS

Charles Robert Darwin (figure 9) was born on February 12, 1809, the
same day as Abraham Lincoln across the Atlantic.[25] He was the second son and the fifth child (of six) of Dr. Robert Darwin of Shrewsbury, a town in the British Midlands (close to the border with Wales),
and his wife the former Susannah Wedgwood. Dr. Darwin's father
was Erasmus Darwin, whom we met in the last section. Like his
father, Robert was Edinburgh educated. Susannah Darwin's father
was Josiah Wedgwood, whom we also met in the last section. A fellow member of the Lunar Society with Erasmus Darwin, he was the
founder of the pottery works that still bears his name, and Susannah's
brother also called Josiah was the father of a brood that included
Emma, Charles's future wife.

At once we can start to place young Charles Darwin. The family was
not aristocratic, but it was middle class, and in the upper echelons.

Both sides of the family were very rich. Robert Darwin was not just a successful physician, he was a powerful money man, using his entrée through his profession into all segments of society and thus able to connect those with money to lend with those with need of money to borrow. Before long, he was lending his own money, and riches accumulated upon riches. The important thing is that this money, and that of the Wedgwoods, was not inherited, a product of land and property long acquired, but very much the spoils of the society of the day, industrial and full of thoughts of Progress for the present and future. Naturally therefore the Darwin-Wedgwood clan fell into the Whig end of the political and social system. Theologically, the Wedgwoods inclined to Unitarianism (believers in God but deniers of the divinity of Christ), and the Darwins were Anglicans (although it is pretty clear that Robert Darwin had little or no belief). Many in the clan felt nothing anomalous about their sympathy for many of the evangelicals' missions, especially an abomination of slavery.[26] Although note the earlier-made point, paradoxically to our generation, being evangelical in emotional commitment almost invariably went along with being a laissez-faire liberal with respect to political economy. No doubt Dickens was exaggerating in *Bleak House* in his portrait of Mrs. Jellyby who spent her days worrying about the poor in Africa, the inhabitants of Borrioboola-Gha, neglecting not just her own family but the wretched of England like Jo the crossing sweeper. But the great novelist wasn't entirely off the mark. Josiah Wedgwood the younger was elected a Whig member of parliament in the early 1830s. Recall that one of its first acts was to put into practice the teaching of Malthus. New Poor Laws were enacted, designed deliberately to make so unpleasant the life of the country's indigent citizens that they would do anything to avoid falling on the mercy of the state. Wedgwood Junior was a "staunch supporter of the ministry," thinking himself in respects almost a Tory, and strongly against the radical end of his party.[27]

The Early Years

Charles Darwin's early life was comfortable and loving and supportive. His mother died when he was young, but the older sisters smoothly picked up the task of child rearing and he seems to have suffered little or no grief at the loss. Above all there was his older brother, also called Erasmus, who acted as a friend and a guide and a source of interest and excitement. At a young age Charles was packed off to one of England's famous public schools (in reality, private schools), Shrewsbury, actually in the town where he lived although like the other boys Charles was a border. He was not a great success at school. He was never that gifted at mathematics, and ancient languages, then a major part of the education, were not things that sparked his imagination or dedication. However going to the school does mark the beginning of what we shall see is the debt of Charles Darwin to his British society and culture. The public schools were very much part of the English establishment, meaning in particular that like the English universities (Oxford and Cambridge) they were at one with the Church of England, the Anglican Church. The headmaster and perhaps some of the senior masters like the teachers at the universities would be ordained ministers, subscription to the basic tenets of the church (the so-called Thirty-Nine Articles) would be obligatory, and religious instruction and church attendance would be part of the daily life. He would therefore be immersed in the rather distinctive nature of the Anglicans. The Bible is very important but so also is the book of nature, reflecting God's power and glory. Natural theology is as much a part of the picture as revealed theology.

Most immediately, we know that from his earliest days Darwin had a love of natural history and collected and horded with passion. A little more formally, what is surely relevant is that he and brother Erasmus became keen amateur chemists, experimenting and devising in a shed in the garden. This is not just one science among equals. It is the science par excellence of industry, especially of a pottery works, where factors like clays and their glazes are of crucial importance. Already we can see that it is a total mistake (often made on the basis of Darwin's misleadingly self-deprecating autobiography) to think that Darwin backed into science casually. From the first he was engaging as a proto-scientist, and the kind of science was dictated by the background—chemistry, on the one hand, and natural history, on the other. The chemistry particularly is fascinating, for it bears far more on our story than one might dream.[28] Darwin was to become one of the greatest theorists in the life sciences, and the texts on which he relied spoke directly to the importance of chemistry for the success of British industry, they stressed the importance of chemistry for the success of the Wedgwood family (whose works was identified explicitly), and they did it all in the metaphors of his class! "The animal body may be regarded as a living machine, that obeys the same laws of motion as are daily exemplified in the production of human art."[29] At the same time, as we shall see with the natural history, the natural theological import was hammered home, for we have "a body of incontrovertible evidence of the wisdom and beneficence of the Deity."[30]

At sixteen, Darwin was sent north to follow in the footsteps of his father and grandfather, to become a physician trained at the University of Edinburgh. This was not a successful move, and two years later he left, bored by the lectures and revolted and somewhat frightened by the operations and general practice of medicine. But he did continue his scientific education—lectures on anatomy and botany and geology (which he found very dull, although he probably got a good grounding in mineralogy)—and then in his second year, much motivated by his general love of science, he fell under the spell of the anatomist and general man of zoology Robert Grant.[31] Stressing, as I do, the fundamental importance of Darwin's British background does not deny the importance of ideas from abroad, most particularly Germany and France. Grant had traveled on the continent, and he knew the latest views about anatomy and taxonomy and the like. In particular, he was fully aware of (and passed on information about) the tensions between comparative anatomists, those like Georges Cuvier who stressed the teleological nature of organisms, their end-directed natures, their exhibition of (what Aristotle had called) "final causes," and those like Etienne Geoffroy Saint-Hilaire, who stressed the holistic nature of organisms, their essential unity with parts being repeated and with analogies (what later came to be called "homologies") between organisms of different species.[32] Function versus Form. Of course, none of this is really a question of British versus non-British. Anyone with a classical education knew that these were issues discussed by the great Greek philosophers, Plato and Aristotle.

Cambridge

Redirected, Darwin turned south toward the University of Cambridge. Following the strong urging of his father, he was now on course toward the life of an ordained minister in the Church of England, and an Oxbridge degree was a necessary condition. In at least two respects Cambridge is very important for our tale. The first is that Darwin was now being exposed to the Anglican ambience full blast. The key here is the works—textbooks really—of the Reverend Archdeacon William Paley. There was the work dealing with revealed theology, *Evidences of Christianity*. This was the work making miracles central to Christian belief. There was also *Natural Theology*, with its famous argument about the watch. This was not one of the set books on the curriculum, but Darwin read it and took its message to heart. Later in life he joked that he could have written it out by heart—except he wasn't joking. Evangelical religion was also a pervading factor. "Repentance, conversion and 'good works' leading to heavenly rewards was the refrain, and nowhere in the Church was it repeated more often that in Cambridge."[33] This was nothing new to the young Darwin, for it is clear that his older sisters (and surrogate mothers) were much inclined this way. The second respect why Cambridge is important is that Darwin's hobbies in natural history started to bloom into full-blown science. Always a collector, Darwin became a fanatic about beetles, building an ever-bigger collection. Of course, in itself, beetle collecting is hardly scientific, but it like everything else to do with the natural world at Cambridge would have been linked with natural theology and thus back to science. The standard work on British insects was the *Introduction to Entomology* by the Reverend William Kirby and William Spence.[34] Right in the introduction one learns that looking at and collecting insects is no purely secular activity. Through nature one gets to God: "no study affords a fairer opportunity of leading the young mind by a natural and pleasing path to the great truths of Religion, and of impressing it with the most lively ideas of the power, wisdom, and goodness of the Creator."[35] As a would-be clergyman, Darwin could justify his hobby with the reflection that its practice was in itself in a sense religious.

This was certainly the theme that Darwin heard incessantly at Cambridge as, during his three-year stay, increasingly he came in contact with senior members of the university who shared his passion for the natural world. These included Adam Sedgwick, professor of geology; William Whewell (figure 6), then professor of mineralogy but to go on to write great works in the history and philosophy of science; and above all John Stevens Henslow, professor of botany. All ordained clergymen, and deeply committed Christians—Sedgwick was an evangelical and Henslow (like Kirby) even went so far as to meddle for a while with then-popular premillennial doctrines about the Second Coming—they knew that not all in their church shared their enthusiasm for science and that some indeed thought it a positive threat (through the denial of the ubiquity of Noah's Flood) to true belief. They all therefore, out of both conviction and expediency, made much of the natural theological implications of their work. Design, design, design would have been the theme drummed into Darwin's ears. It is not science or religion, but science and religion, all focused on seeing the Creator's efforts in play.

Darwin did not take a science degree. There were no science degrees. However, for three years he attended Henslow's botany lectures, so increasingly he was learning about the physiology of plants, about their means of reproduction, as well as about such matters of what we today would call their ecology and their geographical distributions. Classification was also important with Darwin learning that often the way to classify is by digging beneath the immediate useful features, what we would call adaptations and the basis for the design argument, and finding the underlying links, the analogies (homologies) between organisms. This hardly made him a trained botanist, but it did give him a grounding to last all of his life.[36] As with botany, so with geology. At Cambridge, Darwin took no courses in the subject (that is, he did not take Sedgwick's geology lectures), probably (or so he claimed later) because he had been so bored with the subject when at Edinburgh. But he must have picked up some of the excitement of the field from mixing with the older members, and in the summer after his degree was finished he went with Sedgwick on a geological expedition to Wales, having a sort of crash course in the practical management of the subject. And then over and around all of this was discussion about methodology, especially good methodology. The Cambridge men, and others who shared their vision in London and Oxford and Edinburgh and elsewhere, were keen to make a place for professional science in British society and culture. They wanted to show both its intellectual force and its usefulness. To this end, they were eager to articulate the standards and methods of good science, and these were topics much discussed.

This being Britain, and especially at Cambridge, the discussion was dominated by the legacy of Isaac Newton. What did Newton do? What would Newton have us do? One who made a significant contribution to this dialogue was the astronomer John F. W. Herschel (figure 5). In a little book, *A Preliminary Discourse on the Study of Natural Philosophy*, that appeared in Darwin's final year at Cambridge and that (on the strong recommendation of Whewell) he read almost immediately, Herschel laid out a picture of Newtonian science, one that saw understanding essentially as a hierarchical network, with causal laws at the topic acting as premises or axioms for all that follows.[37] From the laws of motion and of gravitational attraction we get Galileo's laws of terrestrial motion and Kepler's laws of celestial motion. Central to this picture is gravity, what Newton called a *vera causa* or true cause. Approaching the question as an empiricist, Herschel argued that when it comes to finding true causes, analogy is the key. Either we spot and experience the cause directly or we experience something analogous. Herschel's example of a genuine *vera causa* is the force attracting the Moon to Earth. We know that there is one because when we spin a stone around at the end of a piece of string, we feel the force down the string as the stone tries to fly off into space. Note incidentally that for Herschel force is the paradigmatic example of a true cause.

The Beagle Voyage

1831! We come now to the big event in Darwin's life. He was not a trained scientist or near anything like this. But he knew a lot more science than many recognize—science steeped very much in the British tradition, and most recently at Cambridge in the British natural theological tradition laced with strong doses of Newton's methodological prescriptions. As importantly, Charles Darwin had been spotted by the Cambridge science elite as a promising new member of the establishment—one with the potential to make major contributions and thereby reflect glory on his old mentors. Through Henslow, Darwin got the offer of a voyage around the world, with much of the time being spent in mapping the coast of South America. The vessel, HMS Beagle, was a British warship commanded by Captain Robert Fitzroy, who was looking for a gentleman companion to ease the lonely hours away from home. It is no idle chance or whim that the Beagle was being directed out to the Southern Hemisphere or that it was the formerly Portuguese and Spanish possessions that were being visited. The Napoleonic Wars had obviously cut off much of the trade with the continent and new markets had been sought. The rich lands across the Atlantic and to the south had beckoned and the encounter had proven highly beneficial. English exports went from virtually nothing when the wars began at the end of the previous century to about 5 million pounds a year at the beginning of the voyage (about 12 percent of the total exports, rather less than the United States but about the same as Asia).[38] Steps had to be taken to ensure that merchantmen do not come to grief in uncharted waters, and it is here that the Admiralty, no longer at war and looking for a purpose, could step in. The Beagle voyage, which was to take five years (figure 11), was all part and parcel of the expansion of British influence that was to dominate the nineteenth century as had no other country.

There is no need to dwell on the familiar details of the story—a story that catapulted Darwin into general fame when his diary-based account of the trip was published commercially to great success.[39] Terribly affected by sea sickness, he did not spend huge amounts of time on the *Beagle* itself, but would make land excursions joining the ship at later periods. Particularly momentous was the time that the *Beagle* spent at the base of the continent, in Tierra del Fuego, returning natives taken to England on a previous trip and trying unsuccessfully to set up a British civilizing mission. Also important were earthquakes experienced in Chile and a visit—one that became much more significant in retrospect—to that group of islands in the Pacific that make up the Galapagos archipelago. But particular events apart, the main theme is the total transformation of the eager but rather naive novice that set off from England into the dedicated professional scientist that returned. Before the voyage, he may have had potential, but he was still a pleasure-loving young Englishman, who would never have thought of putting work before real activities: "at that time I should have thought myself mad to give up the first days of partridge-shooting for geology or any other science."[40] On the voyage, ambition took over, the work patterns of his father and grandfathers asserted themselves, and he became totally dedicated to his mission—to observe, to collect, to preserve, to catalog, to record, to describe. A combination of the training he had had before he left and the training he had gotten on the job, combined with the aid of the really good reference library carried on the ship, meant that the effort was put to good use and that long before he returned (thanks to long, informative letters) Darwin's position in the upper echelons of British science was secure—as it remained until the day of his death. Beyond death actually, given his prestigious burial place.

Geology

Thanks to his authorship of the *Origin of Species*, we think of Darwin as a biologist. In the early years, however, with good reason he thought of himself as a geologist.[41] No surprise therefore that the formative intellectual factor on the *Beagle* voyage was his reading of

Charles Lyell's *Principles of Geology*.[42] First at Edinburgh and then at Cambridge, Darwin had been fed the favored geological theories of the British scientific establishment. Based on eighteenth-century ideas of the German mineralogist Abraham Gottlob Werner, and then reinforced by the thinking of Cuvier, the world was seen dynamically as a series of periods or epochs, broken by times of great upheaval with physical forces, often water-based, having significant effects.[43] By the time of Darwin, this view—that Whewell was to label "catastrophism"—was given a natural interpretation, in the sense that the upheavals were thought law-bound rather than miraculous. What was important was that it was directional, with previous epochs probably much warmer than the present—hence one can account for the existence of fossil palms around Paris—and all coming down to the present. Organisms would be wiped out during the catastrophes, and then new ones would appear when things had again calmed down. This accounts for what Cuvier had shown was the roughly progressive nature of the fossil record. Extinction is for real, but what about the arrival of new organisms? Not much is known here, but unanimous opinion was that here there were indeed miracles. God intervened to create again and again. Not very scientific? Perhaps, but this might be a time when science can say no more. In Whewell's opinion, "science says nothing, but she points upwards."[44]

Darwin had heard of Lyell's ideas before they set sail. Herschel discussed them favorably in his little text, and Sedgwick had given his negative opinion.[45] As they set out, Fitzroy as a present gave Darwin a copy of the first volume of Lyell's work (the subsequent two volumes were sent out as they appeared), and so he was primed when they first put foot on solid soil, the island of San Jago (now known as Santiago, the largest of the Cape Verde islands in the Atlantic). At once Darwin was an ardent Lyellian convert, and this continued through the whole voyage. But what was the geological approach that Lyell, a Scottish-born, English-educated lawyer-turned-geologist (in major part because of bad eyesight) put forward in his work? What was the "uniformitarian" position (as Whewell labeled it) that so attracted Darwin and that was probably the greatest of all influences on his thinking? The full title of Lyell's work gives a clue: *The Prin-*

ciples of Geology, Being an Attempt to Explain the Former Changes of the Earth's Surface, by Reference to Causes Now in Operation. Lyell wanted no part of catastrophes, natural or supernatural.[46] He wanted the kinds of causes that we see around us today—rain, snow, erosion, volcanoes, sedimentation, earthquakes, and the like—to bring about all geological changes, large and small.[47] Mountains, valleys, plains, cliffs, islands, and every other form of covering of this planet are the result of slow but steady regular forces of nature. What this needed of course was time, and this Lyell was willing to grant in abundance. The key phrase was that of the eighteenth-century Scottish geologist James Hutton: "no vestige of a beginning,—no prospect of an end."[48]

There is more to Lyell's theory than just similar causes. They are to be similar causes of the same intensity. No earthquakes of a type and magnitude not previously experienced. No volcanic action unknown to the past or perhaps also to the future. Lyell was prepared to give a flexible interpretation of this rule. At some point, the falls at Niagara will have eaten their way back to Lake Erie, and then there will be a huge emptying of one lake into another. But it all must be within the limits that we today could see happening. And most crucially, everything was to be in a steady state. Lyellian uniformitarianism had no truck with the directionality of the catastrophists. Why this was the case is (on the surface) unclear, because the earlier parts of the program—similarity and intensity of cause—do not seem to dictate that directionality is out of the question. In part Lyell's concerns were empirical. He knew about the problem of the fossil palms near Paris. His reply—his "grand theory of climate"—was that the temperature of a place on the globe is not (as it might seem) a function of the distance from the equator, but more of the relative distributions of land and sea. Citing the Gulf Stream in evidence—an analogy that earned Herschel's enthusiastic approval as exemplary *vera causa* thinking—Lyell argued that it showed that you can get any temperature anywhere if land and sea play their roles properly. The British Isles, judged by comparable places equidistant from the equator, ought to be far colder than they are, but the water crossing the Atlantic from the West Indies means that they are temperate and pleasant. But how do you get redistributions, as has happened since Paris was warm? To use a modern analogy, Lyell suggested that essentially the globe is like a giant water bed with depression in one quarter being balanced by uprising in another. Unlike plate tectonics where things are in constant motion laterally, for Lyell the planet is in constant motion vertically. As mountains rise in one part of the world, the seabed sinks in another.

In part however one senses more of a metaphysical motive driving Lyell. There is no question that Lyell was desperately worried about evolution, and in the second volume of *Principles* gave a detailed exposition and criticism of Lamarck's theory of evolution, albeit tying it to the fossil record as an explanation of progress, something quite absent from Lamarck's own treatment of the issues. Lyell wanted above all to preserve the status of humankind, and evolution was the thin end of a very large wedge making us humans just part of the natural world. But also Lyell liked steady state because it fit with the kind of theological stance that he took. Around this time he was worshipping with the Unitarians and (humans apart) wanted to downplay miracles and an intervening God.[49] A world that exists indefinitely was a world he found theologically congenial. No worries for him about a six-thousand-year-old Earth as calculated by Archbishop Ussher from the genealogies of the Old Testament. (Sedgwick and company would also not have felt so constrained, although they certainly did not want to give the infinite time of Lyell.)

The Disciple

Does this all then suggest that Lyell's thinking had its roots in the industrial culture that absorbed the country in which he lived? It does. Lyell's theory of climate did not come from nowhere. It is a successor of the kind of theory ("vulcanism") endorsed in the century before by James Hutton, where subterranean heat is seen to be the driving factor in geological change, forcing mountains up and then allowing for subsidence as sedimentation and the like did their work.[50] And this in turn did not come from nowhere. It is the natural counterpart to that greatest and most influential of machines of the early Industrial Revolution—the Newcomen engine.[51] Machine and globe both work through heat driving things up then cooling, allowing them to return to their original places. Round and round and round go machine and planet, driven by heat, and ultimately getting nowhere and ending where they began. The influence on Hutton was deep, and it was passed in turn to Lyell. And to Darwin. Remember that he was out in the world, away from his Cambridge mentors, and having to think and geologize for himself. Unlike his mentors he also was a child of the Industrial Revolution, for whom machines were meat and drink. Little wonder that the Lyellian approach was welcome. Darwin had a firm guide to the many and varied phenomena he encountered, and he used it to the full. It is true that there was some modification along the way. Lyell seems to have thought of land popping up or down a bit like a whack-a-mole. Darwin seems to have thought of the rises and falls as bigger and more sedate. South America is on its way up; the Pacific floor is on its way down. But the picture and the mechanism are the same.

This is not an intellectual biography so there is no need to cover the work in detail. Ask rather about the religious beliefs of this young Lyell convert. Simply, from a convinced Anglican at the beginning of the voyage Darwin's thinking changed naturally and gradually to a kind of deism, the religion of his mother's family. God was an unmoved mover, and miracles were no longer acceptable. As we have seen all of this would have been interconnected and smoothly leading from one position to another. A religion that puts so much trust in miracles is vulnerable when they start to go, as they did for Darwin both as he pondered the unlikeliness of the biblical accounts and as Lyellian geology convinced him that they are unnecessary to explain the natural world.

> By further reflecting that the clearest evidence would be requisite to make any sane man believe in the miracles by which Christianity is supported,—that the more we know of the fixed laws of nature the more incredible do miracles become,—that the men at that time were ignorant and credulous to a degree almost incomprehensible by us,—that the Gospels cannot be proved to have been written simultaneously with the events,—that they differ in many important details,

far too important as it seemed to me to be admitted as the usual in-
accuracies of eye-witnesses;—by such reflections as these, which I
give not as having the least novelty or value, but as they influenced me,
I gradually came to disbelieve in Christianity as a divine revelation.[52]

Nor, as we shall see later, was humankind to be a big stumbling block
for Darwin. To the end of his life, Lyell agonized about our species.
He was never quite able to "go the whole orang."[53] Darwin was never
like this. Perhaps it was part of his temperament—"I do not think
that the religious sentiment was ever strongly developed in me."[54]
Perhaps it was in part his experiences. A man who had seen the Tierra
del Fuegians in their natural state was never going to argue for our
ontological or theological distinct status. So increasingly by the fall
of 1836, when the *Beagle* returned to England, we have a young man
perfectly comfortable with a God who set all in motion and then let
events take their course. The God of the British industrialist. This was
not a matter of any great tension for Darwin. Like the great wealth, it
was part of his heritage—especially as he moved from the evangelical influences of his early years to the thinking of his more skeptical brother and father. Machines, factories, industry, political economy, Progress. The God of deism: the God of people who think that
God is an unmoved mover, who works through unbroken law rather
than through direct interventions (the God of theism). This is not an
absent God. This is as much a creator, designer God as the Christian
God. And if you take someone like Herschel seriously, this is a God
who is present in His laws as they work their magic. It is a God of
self-control and great forethought. Notice also how this makes God
Himself an industrialist, where things are done through machines
rather than by hand—the power loom rather than the hand loom.
"Precisely in proportion as a fabric manufactured by machinery affords a higher proof of intellect than one produced by hand; so a
world evolved by a long train of orderly disposed physical causes is a
higher proof of Supreme intelligence than one in whose structure we
can trace no indications of such progressive action."[55]

And so now we start to build up to the big question. In the next two
years, Darwin is to become an evolutionist and to find his mecha-

nism of natural selection. How did this happen, and in particular how is this related to and explained by his very British upbringing and modes of thought?

EVOLUTION AND NATURAL SELECTION

Becoming an Evolutionist

Charles Darwin knew about evolution from a very early age. Apparently even as a teenager he read parts of his grandfather's evolutionary tract, *Zoonomia*. Then at Edinburgh there was the encounter with Robert Grant, who of all people was about the one person in Britain openly to endorse evolutionary ideas that he had picked up from his time in France. After that, there was Lyell's very detailed exposition and discussion of Lamarck, so detailed in fact that later Herbert Spencer was not alone in being converted to evolution by Lyell's treatment! There is no reason to think that Darwin fell into this category, but on the *Beagle* voyage he must have wondered about the origins of organisms. Even if Darwin had not wondered, others certainly were puzzled. Herschel, in the mid-1830s down in South Africa mapping the southern skies and whom Darwin met toward the end of the *Beagle* voyage, wrote to Lyell about precisely this problem, what he famously called the "mystery of mysteries," affirming his belief that the solution had to be one that was natural, that is, that was subject to regular laws.[56] The problem was out there to be solved and must have attracted a bright and ambitious new entrant to the scientific community. An entrant, moreover, who now had more direct knowledge about a lot of the pertinent information than anyone else. Darwin had (on the voyage) been puzzled by the fossil formations that he saw, where forms very similar take the places of older forms. He had noted with interest and surprise the way in which one kind of organism will supplant another kind as one moves sideways geographically. And then there was the visit to the Galapagos. There was no eureka moment, but even before the voyage was ended he was thinking hard. Geographical distributions were absolutely fundamental to a Lyellian geologist. By looking at the extant flora and fauna you

can make inferences about the past. Darwin had done this most successfully when he compared the organisms on the two sides of the
Andes and, finding them similar, reasoned that although massive the
mountains are recent. The Galapagos was incredibly intriguing, because there from island to island one found somewhat different reptiles and birds. Why was this?

It is almost certain that the answer did not come until some six
months after the voyage, when Darwin had given his birds to a com-
petent taxonomist to study, and when the birds from different islands
were declared different species. There could be no other answer.
They had transformed naturally from a single parent stock. Evolution had occurred! Was there any tension? One really does not sense
that there was or that, the move once made, Darwin ever thought
once of backtracking. No more should there have been. Once you
take out of the picture Lyell's obsession with the status of humans,
which immediately eliminates the need to see the fossil record as
steady state, then (as Herschel noted) you are on the road to a natural
solution, and what else is there but evolution? It is hardly plausible to
think that organisms are created anew from mud, rather like a law-
bound version of the early chapters of Genesis. If you add to this that,
far from being antireligious, you are providing precisely the kind of
answer you would expect given your deistic God, then the science
and the faith blend happily together. And if this were not enough, on
his return from the voyage, Charles Darwin was much in the company of his older brother Erasmus, who was now set up in London as
a man about town. He was living on the family money, having neither
employment nor need or desire of such a burden, mixing in radical
literary circles that included the sage Thomas Carlyle and his wife,
Jane; Harriet Martineau, political theorist, sociologist, essayist, radical, and much more; and Charles Babbage who was spending huge
amounts of public money in his attempt to build a calculating machine, a kind of proto-computer. Although Babbage was never successful in his attempts, he showed that in theory all sorts of wonderful sequences were possible, and that these would occur entirely
according to unbroken law.[57] A machine could be programmed
to give a million numbers, in order, and then suddenly start to go

haywire giving, instead of the expected 1,000,001 the unexpected 1,000,010. Or it could be made to give any other number instead. In short, generalizing, there is no need to think of an unexpected event as caused by divine interposition, as a miracle. The sequence could all have been determined in the first place.

The God of the Machine. The God of Industry. The idea of evolution raised no tension for Darwin nor did its acceptance. There was however tension about how he was to deal with his new conviction. Back in England, he was the rising star, cherished and admired by the members of the senior establishment, who with some good reason were priding themselves on finding and encouraging so brilliant a new entrant into the ranks of professional science. Even people like Lyell and Herschel were uncomfortable with evolution, and the more conservatively theologically inclined, notably the Cambridge professors, were violently against it. This was to come out in full force in 1844 when the *Vestiges of the Natural History of Creation* appeared on the scene. The reactions were, to say the least, extreme.[58] Best of all was Sedgwick, who followed up a violent review (1845) by republishing a little thirty-page sermon with a five-hundred-page foreword and a three-hundred-page afterword (1850) on the iniquities of evolutionary thought![59] So Darwin had to keep his ideas to himself; but in no way did this deter him. After five years on the *Beagle*, he was his own man.

Becoming a Darwinian

Thus far we have seen the fruit of the one British side of Darwin's inheritance. His family background—the industrialist, deistic, radical thought—made him an evolutionist. This however is only half of the story. We are now going to see how the Anglican side made him a Darwinian. For a start, why didn't Darwin just stop where he was? At least, why not remain content as an evolutionist and work hard to make it plausible? Darwin was already a geologist and a good one. Naturally therefore he had interests in paleontology. Why didn't he devote the next few years to the fossil record, looking at the discoveries that he had made as well as the many discoveries that were being made by others in Europe and elsewhere? Why didn't he build the empirical evidence for evolution—the fossils, the geographical distributions, and more—to the full? Clearly this was a job that was going to have to be done at some point. But Darwin was little tempted in this direction. He was an Englishman. He was a Cantabrigian. Above all, he was a Newtonian. Whewell had just published his magisterial three-volume *History of the Inductive Sciences*. Darwin devoured it, twice before the year was out, marking up in a highly suggestive manner the achievements of the great physicist.[60] That meant causes. It meant *verae causae*—true causes. Immanuel Kant said that there could never be a Newton of the blade of grass. Darwin was determined to show him wrong. So the hunt was on for the cause of evolutionary change. Notebooks were purchased, covers were marked "private," and the young scientist was set to speculate.

> Astronomers might formerly have said that God ordered each planet to move in its particular destiny.—In same manner God orders each animal created with certain form in certain country, but how much more simple & sublime powers let attraction act according to certain law such are inevitable consequen[ces].
>
> Let animals be created, then by the fixed laws of generation, such will be their successors.—Let the powers of transportal be such, & so will be the forms of one country to another.—Let geological changes go at such a rate, so will be the number & distribution of the species!![61]

Kant (no reason to think that Darwin read him at this stage although Whewell might have passed on some of the ideas) is a good guide at this point. He himself in the *Third Critique* toyed with evolutionary ideas.[62] Most impressive were the analogies (homologies) between organisms. Could these be just chance, or did they point to something deeper? Many in the German Romantic movement, looking for a neo-Platonic unity throughout existence, thought the analogies all-significant. It seems clear that Goethe among others was attracted to evolution for this very reason. But as with Kant himself, this could never be enough for Darwin. The key fact about organisms is that they are design-like, they exhibit final causes. For Kant, as for the German-educated, Kant-influenced Cuvier after him, you simply have to account for this special organization—what Cuvier called the "conditions of existence"—and blind law is not enough. This of course fit smoothly with the obsessions of British natural theology, and the Cambridge set took up Cuvier as a strong scientific support of what they already believed theologically. So Darwin—unlike Chambers who was blithely to ignore or dismiss the issue of final causes—knew that any satisfactory causal answer had to explain design, adaptation.

A lot of heady speculation went into the early days of Darwin's evolutionism. Looking at the notebooks, there are many suggestive ideas although one can never be entirely sure about what he was thinking—one suspects that he was not that sure himself. But there was a guide, one supplied by Herschel and exemplified by Lyell. Look for analogies. You might ask why Darwin never looked to see causes actually working today—working today so powerfully that one can see the causes and the effects at once or at least in a reasonable amount of time. To the end, Darwin never thought that way, always assuming that change will be so slow as to be unobservable. The Lyellian spell demanding huge amounts of time was at work here. But if analogies, where do we find them? Here the family background came to the fore to speak to the questions and demands of the Cambridge education. As we know, an industrial revolution presupposes an agricultural revolution. You cannot pack a growing population off to cities unless you are able dramatically to improve supplies of food and drink. You need more cows, more vegetables, more hops, and more barley if you are to feed the workers and quench their thirsts. We know also that the key here is selective breeding—take your best cows and sheep, your best turnips and raspberries and hops and barley, and use them as the stock for the next generation. As so frequently the case in this tale, Darwin was in an ideal position to find and use this knowledge. He came from the heart of rural England, where animal and plant breeding was a central activity, and from a family that was itself trying its hand at the practice. His uncle Josh (to become his father-in-law early in 1839) had set himself up as a kind of gentleman farmer, and was experimenting at introducing new sheep breeds into Britain.

It did not take much to see that in the improvement lay the key to the problem of final causes. Farmers—and also we must include pigeon fanciers and dog breeders and racehorse owners and the like—did not breed just by chance (figure 18). They were out for improvement. Fatter pigs and shaggier sheep and fleshier turnips. They were in the business of designing for useful ends. But does this, can this, tell us about nature itself? Slowly but surely Darwin convinced himself that this is so. He even read a pamphlet that anticipated his own great discovery suggesting that nature can step in where humans do not tread.[63]

> A severe winter, or a scarcity of food, by destroying the weak or unhealthy, has all the good effects of the most skilful selection. In cold and barren countries no animal can live to the age of maturity, but those who have strong constitutions; the weak and the unhealthy do not live to propagate their infirmities, as is too often the case with our domestic animals. To this I attribute the peculiar hardiness of the horses, cattle, and sheep, bred in mountainous countries, more than their having been inured to the severity of climate.[64]

Darwin took careful note of this passage, and even though he could not quite see the full import grasped that if something like this went on long enough, we would get full-blooded species.

Then, famously, at the end of September 1838, some eighteen months after he had set out to find a cause, Darwin read the sixth edition of *An Essay on a Principle of Population* by Malthus. Although he did not seek out the work, it was hardly chance that someone like Darwin, systematically searching the literature, looking for clues to solve his puzzle, should have read Malthus. Given the literary and cultural interests of Darwin's older brother Erasmus (whose copy of Malthus it was that Darwin did read) and given how the Darwin-Wedgwood family took Malthus's ideas as virtually true a priori—certainly comforting support for its attitude toward its own factory workers in the pottery works—Darwin's encounter was nigh inevitable. In fact, in respects it was refreshing thinking known already, because in the all-important *Principles*, Lyell made references to the "struggle for existence." In his reading of *Essay*, Darwin would have seen how Malthus had taken his human-centered ideas from speculations by Benjamin Franklin about the animal world. It was easy to apply them back again. Moreover, you should not think that here the Anglican side to Darwin's influences would have been much opposed to this Whig credo. The Cambridge set would have been at one with Malthus in putting everything in a natural theological context. All would have agreed that we are dealing with God's laws and that He put them in place to prevent human complacency. He would not want us simply doing nothing with our lives, except eating, drinking, and copulating. There have to be pressures to make us work and try to create something of our situation. The argument was about the creation of something and not about the pressures as ends in themselves.

For Darwin, all fell into place. The Malthusian laws created a pressure, and in nature this means that some survive and reproduce and others do not. We have a natural equivalent of the breeders' art of selection, for the success in the struggle for existence is not random but a function of the different features that are possessed by the actors in the drama. And this means we have an answer to the problem of final causes. The end result is going to be change but change in the direction of design-like features.

Population is increase at geometrical ratio in far shorter time than 25 years—yet until the one sentence of Malthus no one clearly perceived the great check amongst men.—there is spring, like food used for other purposes as wheat for making brandy.—Even *a few* years plenty, makes population in Men increase & an *ordinary* crop causes a dearth. take Europe on an average every species must have same number killed year with year by hawks, by cold &c.—even one species of hawk decreasing in number must affect instantaneously all the rest.—The final cause of all this wedging, must be to sort out proper structure, & adapt it to changes.—to do that for form, which Malthus shows is the final effect (by means however of volition) of this populousness on the energy of man. One may say there is a force like a hundred thousand wedges trying force ~~into~~ every kind of adapted structure into the gaps ~~of~~ in the oeconomy of nature, or rather forming gaps by thrusting out weaker ones. —[65]

Note the language of "force." What Darwin saw himself as having here is a cause that is a force—a kind of strength or power that is exerted on things—the very thing that Newton posited at the heart of his great theory of the physical sciences.

No one grasps a discovery immediately and fully. But within a month or two Darwin knew what he was about and was seeing how it could be applied. Indeed, the first clear statement of what he was to call "natural selection" came later in the year and was applied to, of all species, human beings, and not just to our physical features but to our mental attributes.

An habitual action must some way affect the brain in a manner which can be transmitted.—this is analogous to a blacksmith having children with strong arms.—The other principle of those children. which *chance?* produced with strong arms, outliving the weaker one, may be applicable to the formation of instincts, independently of habits. —[66]

The most obvious thing about this passage is how firmly Darwin was committed to what has come to be known as "Lamarckism," the inheritance of acquired characteristics. He always accepted this as a secondary mechanism, and there is reason to think that later in life he came to rely on it more. But more significant is the way the passage shows that Darwin saw that the creative factor in selection must come from the selecting itself. It cannot be aided by built-in direction of the organic features on which it acts. That for Darwin would be to introduce into the world a direction, a teleology, that it was the whole purpose of his mechanism to eliminate or at least to render redundant. We will be returning to this point.

Charles Darwin was a genius. Let there be no mistake about that. Moreover, he was a hardworking genius. His ideas did not just float into being, unbeckoned and without effort. Systematically he set about his science, and when challenged by the biggest puzzle of them all, he went out and sought the solution and with much effort found it. But none of this is to deny the background. It is rather to confirm it. He was lucky, not in his inferences—those were his and his alone—but in his influences. Few could have had the very inputs from across the spectrum that were received by the younger son of Dr. Robert Darwin, and it was these inputs that he put to good use. Darwin was a revolutionary, but his genius and his achievement was not that of a Good God, making things out of nothing. He took the pieces that he had, and like a children's puzzle he rearranged them to make something else. Now the pieces say one thing. Now they say something else. And the all-important point is that these pieces were drawn from his British heritage. Darwin's was a theory that sprang from the mechanized, industrialized, political-economic society of the late eighteenth and early nineteenth century uniquely to be found on the little island on which he lived—a theory that was spliced with the Anglican-theological and Newton-admiring educational society that had welcomed him in and that was making him one of the brightest stars in their firmament.

Making a Theory

This influence was to continue, for a mechanism of change is not a theory. Herschel and increasingly Whewell were laying out the conditions of the best kind of theory, the best kind of Newtonian theory that is. After Whewell had written and published his massive *History*, three years later in 1840 he followed it up with a further two volumes of the *Philosophy of the Inductive Sciences*.[67] I am not sure that Darwin ever did read the *Philosophy*, but he did read Herschel's very detailed and lengthy review in the *Quarterly Review* in 1841, responding with enthusiasm.[68] And of course he had been listening to Whewell since his days as an undergraduate at Cambridge.

The basic structure of a good theory had to be "hypothetico-deductive," that is, a kind of empirical law network or axiom system, with high-powered laws at the top and more phenomenal laws at the bottom. And it had to be causal, just like Newton's theory of mechanics. Herschel wanted analogies to show that one has *verae causae*, and we have seen Darwin accept this fully. However, Whewell had a rather different take on things. Always inclined to be more of a rationalist to Herschel's empiricism, Whewell worried that Herschellian *verae causae*, exemplified as we have seen by Lyellian geology, automatically favored uniformitarianism over catastrophism. The whole point of the latter approach is that we do not see or otherwise experience the massive upheavals. There is no analogy today. So Whewell had to find a *vera causa* principle of his own, and it was here that his celebrated "consilience of inductions" came into play.[69] (I say "his" because Whewell highlighted it, but the idea can be found in the writings of others, including Herschel.) A *vera causa* is something that unites a number of disparate areas of study, under one hypothesis. Newtonian gravity is a *vera causa* because it brings together terrestrial and celestial physics. Likewise catastrophes are *verae causae* because they explain the various facts of geology.

One should say that Whewell was not simply arguing to preserve a history that meshed nicely with his theological interests: he had other important arrows in his quiver. The first part of the nineteenth century saw an intense debate about the nature of light, with the balance swinging firmly from the particulate theory of Newton to the wave theory of Huygens. The experiments of people like Young and Fresnel put the matter beyond debate. But no one sees the waves! How then can they be *verae causae*? Herschel tied himself into knots on this one, with all sorts of analogies about tuning forks and bits of string and sealing wax, trying to show how sound can be the desired analogy.[70] Whewell just laughed this off and argued that waves stand at the center of a consilience and that is all that is needed.[71] It is pertinent to note that one of his young listeners took this to heart and whenever he was challenged that no one saw his mechanism in action he at once made reference to the wave theory of light.[72] Darwin knew a good point when he saw one.

Darwin sketched his theory in 1842 and then wrote it out more fully in 1844.[73] He now had a structure that remained unchanged right through to and including the *Origin*. Mistakenly, as we would now judge, Darwin thought there could be no direct evidence of evolution through natural selection. So he went about showing that it is a *vera causa* because it satisfied both the Herschellian empiricist criteria and then the Whewellian rationalist criteria. He was hardly able to embed his thinking in anything as formal as the Newtonian system—the information was not there and the mathematical ability was lacking—but he did then (and always) try to show that natural selection did not just appear but was a consequence of general laws of nature.[74]

First, there is the inference to the struggle for existence.

> De Candolle, in an eloquent passage, has declared that all nature is at war, one organism with another, or with external nature. Seeing the contented face of nature, this may at first be well doubted; but reflection will inevitably prove it is too true. The war, however, is not constant, but only recurrent in a slight degree at short periods and more severely at occasional more distant periods; and hence its effects are easily overlooked. It is the doctrine of Malthus applied in most cases with ten-fold force.[75]

"The doctrine of Malthus": population numbers go up geometrically and food and space are confined arithmetically. The result is a force, a pressure, never ending: "Nature may be compared to a surface, on which rest ten thousand sharp wedges touching each other and driven inwards by incessant blows."

Then on to selection. There will be ongoing natural variation for whatever reason—Darwin always thought of this as "chance," not in the sense of being uncaused but of our being ignorant of the causes and (most importantly) not appearing according to need, being "random" in some wise. Then, given this variation, the struggle takes over.

> Now can it be doubted from the struggle each individual (or its parents) has to obtain subsistence that any minute variation in structure, habits, or instincts, adapting that individual better to the new conditions, would tell upon its vigour and health? In the struggle it would have a better chance of surviving, and those of its offspring which inherited the variation, let it be ever so slight, would have a better

chance to survive. Yearly more are bred than can survive; the smallest grain in the balance, in the long run, must tell on which death shall fall, and which shall survive. Let this work of selection, on the one hand, and death on the other, go on for a thousand generations; who would pretend to affirm that it would produce no effect, when we remember what in a few years Bakewell effected in cattle and Western in sheep, by this identical principle of selection.[76]

Note the final reference to the breeding of cattle and sheep using selection. Before these key passages, Darwin had already introduced this analogy to the reader, trying to show how such breeding starts with original stock and then divides and diverges and changes, in a direction of utility. In part, no doubt this is heuristic, as well as autobiographical. But it is also intended as part of the evidentiary argument. Given what goes on in the farmyard, it is reasonable to think something similar goes on in nature. Then this done, Darwin was free to turn to offering a Whewellian consilience, which now took over the discussion. And so such topics as instinct were introduced, followed by paleontology, geographical distributions, anatomy and morphology, classification, and embryology. Darwin went through the life sciences showing that the phenomena encountered can be explained by evolution through natural selection and conversely how the explanations increase the reasonableness of the causal hypothesis of evolution through natural selection.

Why, for instance, are the embryos of organisms very different as adults so often so similar? Because variations can appear at any time in an organism's development and because selection would be working to divide the adults not the embryos. And at this point the clever Darwin circled back and used the Herschellian approach to justify his case.

Whatever may be thought of the facts on which this reasoning is grounded, it shows how the embryos and young of different species might come to remain less changed than their mature parents; and practically we find that the young of our domestic animals, though differing, differ less than their full-grown, parents. Thus if we look at the young puppies of the greyhound and bulldog—(the two most obviously modified of the breeds of dog)—we find their puppies at the age of six days with legs and noses (the latter measured from the eyes to the tip) of the same length; though in the proportional thicknesses and general appearance of these parts there is a great difference.[77]

It is not my aim here to argue whether Darwin was right or wrong. I am simply trying to show what he was up to and why. And bringing this part of the discussion to an end, let me tie it in again with the British situation in which Darwin was embedded. The structure of Darwin's theory was set up to fit certain specified norms of scientific excellence. These norms did not come from nowhere nor was it chance that Darwin was so keen to satisfy them. He wanted to produce good science by the standards of the day—by the standards articulated by the leaders of the day—even though (from a formal viewpoint) he could not always measure up to this science. But the norms were not just epistemological. They were social. Paradoxically although England was ahead of the world on industry and technology, it was behind on pure science and its support, something that came to haunt the country later in the century. The Cambridge network and friends were trying to upgrade the status of science and of its practitioners, something not easy in Britain as opposed to France and Germany. To this end they tried to reform the Royal Society, they founded the British Association for the Advancement of Science, they sought government support for the leaders of the field (like Richard Owen), and more.[78] And part of the "more" was to spell out the nature of good science. Hence, the writings of Herschel and Whewell and others (including Baden Powell in Oxford). Sociologically, Darwin's was a very British theory.

ON THE ORIGIN OF SPECIES

Charles Darwin wrote out his full account of his theory in 1844. And then was silent for fifteen years. He was married to his first cousin Emma. The young couple had moved to seclusion in Kent, where they were raising an ever-growing, Victorian-size family. Darwin had also fallen sick of an illness that to this day remains a mystery—the latest hypothesis is that it might have been lactose intolerance.[79] Using this malady to a certain extent as a tool to avoid society—compensated by the new efficient mail service (the "penny post") made possible by the spread of the railways, thus keeping him in close touch with fellow scientists and knowledgeable correspondents, who together functioned as a kind of Internet encyclopedia a hundred and fifty years before its time—Darwin buried himself in a massive exercise in barnacle taxonomy that dragged on and on. Basically he was terrified of publishing for he would have been anathema in the group to which he belonged and of which he was one of the brightest stars. Finally, in the mid-1850s, he returned to evolutionary work and at the same time, to replace the older network, built around him a group of younger workers who would embrace and carry forward the ideas of evolution. Then, as the world knows, the young naturalist Alfred Russel Wallace sent to Darwin an essay with the very ideas of evolution through natural selection, and thus prodded Darwin wrote the *Origin of Species*, a work published toward the end of 1859.[80]

The overall structure of the *Origin* is that of the original essays—artificial selection, natural selection, and then application across a broad field, the consilience. As commentators have noted, Darwin does not separate out the fact of evolution from its cause, and throughout the volume the argument is for evolution for all organisms, brought about primarily by natural selection.

Natural Selection

For Darwin natural selection is not just a cause, but a force in a kind of Newtonian sense. It is this that molds the living world to its ends. "It may be said that natural selection is daily and hourly scrutinizing, throughout the world, every variation, even the slightest; rejecting that which is bad, preserving and adding up all that is good; silently and insensibly working, whenever and wherever opportunity offers, at the improvement of each organic being in relation to its organic and inorganic conditions of life."[81] To the best of my knowledge, although Darwin freely uses the word "mechanism" of contrivances, that is, complex adaptations that are machinelike, he never speaks of selection as a mechanism.[82] Although common today, it seems to be a practice that came only at the beginning of the twentieth century. This is not terribly significant; the force of selection is located within the overall machine metaphor, making mechanisms and in turn being powered by mechanisms. For this very reason, that machines are built for a purpose, note that selection is most crucially a force that speaks to final cause. For Darwin—as for Paley—this is the defining fact about animals and plants. By the mid-1850s, homology (as we can now call it without anachronism) was being seen as the crucial issue in the living world. It was all-important for leading morphologists like Richard Owen, who had a theory of archetypes to explain the cross-species isomorphisms.[83] For all their differences, this was the view of younger biologists like Thomas Henry Huxley.[84] Even Whewell, tying himself into metaphysical knots about the inhabitants of other worlds, was now stressing homology over adaptation.[85] And Darwin of course had to use homology extensively in his work on barnacles.[86] But the Darwin of the late 1850s was still the Darwin of the 1830s, when final cause was all-important. And this was the message of the *Origin*. There is "Unity of Type" (homology) and "Conditions of Existence" (adaptation), and let there be no mistake about the priority.

> On my theory, unity of type is explained by unity of descent. The expression of conditions of existence, so often insisted on by the illustrious Cuvier, is fully embraced by the principle of natural selection. For natural selection acts by either now adapting the varying parts of each being to its organic and inorganic conditions of life; or by having adapted them during long-past periods of time: the adaptations being aided in some cases by use and disuse, being slightly affected by the direct action of the external conditions of life, and being in all cases subjected to the several laws of growth. Hence, in fact, the law of the Conditions of Existence is the higher law; as it includes, through the inheritance of former adaptations, that of Unity of Type.[87]

Isn't there something a bit fishy about all of this? Even if we grant that Darwin can explain final cause naturally—meaning that he can explain final cause without special guided laws or whatever—is it really the case that natural selection itself is so very natural? Back in the earliest versions of the theory (the early 1840s), the analogy with human selection was made so strongly that it is hard not see natural selection as God's direct enterprise, something He was really involved in and very far from the mechanism of an unmoved mover.

> Let us now suppose a Being with penetration sufficient to perceive differences in the outer and innermost organization quite imperceptible to man, and with forethought extending over future centuries to watch with unerring care and select for any object the offspring of an organism produced under the foregoing circumstances; I can see no conceivable reason why he could not form a new race (or several were he to separate the stock of the original organism and work on several islands) adapted to new ends.[88]

And on and on it goes, with the Being's abilities being "incomparably greater than those qualities in man," and the end results "greater than in the domestic races produced by man's agency." Indeed: "With time enough, such a Being might rationally (without some unknown law opposed him) aim at almost any result."

Two comments are in order. First, Darwin—certainly by the time of the *Origin*—was speaking analogically or metaphorically. He was trying to say what is happening and the anthropomorphism was no more essential than if we say something like "the eye is incredibly well designed" or "molecular biologists quickly cracked the genetic code." In a later edition of the *Origin*, Darwin made this point himself. "It has been said that I speak of natural selection as an active power or Deity; but who objects to an author speaking of the attraction of gravity as ruling the movements of the planets? Every one knows what is meant and is implied by such metaphorical expressions; and they are almost necessary for brevity." He stressed that "I mean by Nature, only the aggregate action and product of many natural laws, and by laws the sequence of events as ascertained by us."[89]

Second, there is something to the charge that selection in the *Origin* is more intentional—even God-infused—than one expects in the cold, hard world of Cartesian *res extensa*, material substance. In part this is because the eye is design-like in a way that the moon, for example, is not. So the metaphor of design continues to be appropriate in Darwinian biology in a way that is not true of physics. Darwin never refers to Kant on this matter, but he would surely have agreed with the author of the Third Critique (the *Critique of Judgment*) that you cannot do biology without the metaphor. The point is that there is no implication of an Aristotelian vital force objectively out there in nature making for final causes. Final causes are our way of thinking about a mechanistic system. In part the intentionality of selection comes because, as always, remember that we are rooted in the 1830s. Herschel was explicit—and Whewell would have agreed—that force ultimately is a matter of God's will. Nothing happens save through His powers. This is the key to the Anglican natural theology of the day. God is not some absent entity—like Aristotle's unmoved mover who spends its time contemplating its own perfection and has no interest in, probably no knowledge of, humankind. Darwin absorbed this aspect of Anglicanism, and we know that down to and including the *Origin* this infused his deistic theology.[90] The many references to the Creator in the *Origin* are not idle. Darwin did not want God intervening directly—that is why, when his American supporter Asa Gray argued for directed variations, Darwin would have nothing of them[91]—but God is nevertheless working His purpose out. Darwin made this clear in a celebrated letter to Gray. "I am inclined to look at everything as resulting from designed laws, with the details, whether good or bad, left to the working out of what we may call chance." Elaborating: "I can see no reason, why a man, or other animal, may not have been aboriginally produced by . . . laws; & that all these laws may have been expressly designed by an omniscient Creator, who foresaw every future event & consequence."[92] As a scientist, Darwin wanted to keep God out of the story. But theologically, the Anglican God was right there at work.[93]

The Adam Smith Influence

Move now to some issues surrounding this notion of natural selection. I want first to note how, starting with the key notions of struggle and selection themselves, Darwin embedded his discussion in the practices, events, and metaphors of his day, his British industrial culture. Looking back, we tend to think of the vigor, the success of British society; we think of how so small a country could push forward with such results. But there was the dark underbelly, the ferocious competition with failure being the norm. There was, for example, a major financial crisis in 1825 and 1826.[94] The failure of a major London bank at the end of the first year brought down forty-three corresponding county banks, and another eighty followed in the early months of the next year. Of 624 companies formed in 1824–25, five hundred had gone by 1827, and only fifteen could be called reasonably successful. A child of industrialists, however ahead of the game, had always before him the struggle, the striving, the sometime success, the usual failure. A popular view was that things would even themselves out, a "balance of nature." Darwin kept that happy illusion at arms-length.[95] He agreed that we can get an equilibrium, one that sometimes lasts, but always there is the threat of things going haywire. Slapping down even Lyell on this matter, Darwin stressed that the focus must always be on struggle.

But remember that success and failure come not just by chance, but are functions of superior or inferior qualities and practices. Here too Darwin drew on his background. Take the "division of labour," that idea that goes back to Adam Smith who stressed that, by dividing up the tasks, much more could be produced by far fewer.[96] It is one of the key notions of the Industrial Revolution, and as a metaphor it is of crucial importance to Darwin's thinking about the workings of selection.[97] First, he wanted to apply it to selection working on individuals. Talking of the adaptations of plants and of the origin of sexuality, Darwin seized on the pollination of plants by insects: "as soon as the plant had been rendered so highly attractive to insects that pollen was regularly carried from flower to flower, another process might commence. No naturalist doubts the advantage of what has been called the 'physiological division of labour'; hence we may believe that it would be advantageous to a plant to produce stamens alone in one flower or on one whole plant, and pistils alone in another flower or on another plant."[98] Working on slight variations affecting fertility, in time "as a more complete separation of the sexes of our plant would be advantageous on the principle of the division of labour, individuals with this tendency more and more increased, would be continually favoured or selected, until at last a complete separation of the sexes would be effected."[99]

Then Darwin moved on to the crucial issue of how the pattern of evolution is essentially and deeply treelike. This had always been his vision and in the early notebooks there are some prescient speculations about causes, but Darwin himself always dated the real discovery of his "Principle of Divergence" to around 1850. And the division of labor was crucial.

> The advantage of diversification in the inhabitants of the same region is, in fact, the same as that of the physiological division of labour in the organs of the same individual body. . . . No physiologist doubts that a stomach by being adapted to digest vegetable matter alone, or flesh alone, draws most nutriment from these substances. So in the general economy of any land, the more widely and perfectly the animals and plants are diversified for different habits of life, so will a greater number of individuals be capable of their supporting themselves.[100]

It is hard to imagine a more British approach to nature than this.[101] Although it is paralleled by Darwin's treatment of another thorny issue, what has come to be known as the "levels of selection" problem. Natural selection is fueled by the struggle for existence, but who is struggling with whom? The answer to this question then tells us about adaptations. Who benefits? On whose behalf is the struggle and, thus, what is the end result of selection? Darwin was adamant that adaptations can never cross species boundaries in the sense that selection works against an individual and in favor of the members of other species. It may well be that one can benefit from other species—the bee gets honey from flowers—but the end of selection is within the species. The bee benefits from the flowers; the flowers benefit from pollination. No one is doing anyone any favors. But what about within the species? Obviously some, perhaps most, adaptations benefit the individual, but could adaptations ever benefit other species members, perhaps at the individual's cost?

Using the obvious terms "individual selection" and "group selection" (not Darwin's language) prima facie the prospects do not seem promising for group selection. The struggle is between individuals within and without the species: "as more individuals are produced than can possibly survive, there must in every case be a struggle for existence, either one individual with another of the same species, or with the individuals of distinct species, or with the physical conditions of life."[102] Darwin even went on to stress that within-species struggle is where the real tensions occur: "the struggle almost invariably will be most severe between the individuals of the same species, for they frequent the same districts, require the same food, and are exposed to the same dangers."[103]

There is also the matter of Darwin's secondary selection mechanism, sexual selection. Divided into two kinds, male combat and female choice, the individual-selection-based nature of these processes needs no argument. Males are fighting it out for access to the females, and the winners have lots of offspring and the losers few or none. Females are choosing the showiest males and again the winners have lots of offspring and the losers have none. There is nothing in it for the group here and everything for the individual. Indeed, one might argue that from the group perspective having a bunch of testosterone-fueled males, always on the rampage against each other, is a recipe for disaster. Likewise having a bunch of males with tail feathers so long and gross that they can hardly move is no great way of preserving the species. But from the individual perspective, it is all important.

As we understand Darwin's treatment of divergence in terms of the metaphors of his day and class—there we find the division of labor—so we naturally understand Darwin's treatment of the level of selection in terms of the metaphors and claims of his day and class. "It is not from the benevolence of the butcher, the brewer, or the baker that we expect our dinner, but from their regard to their own interest." Anything organisms do for others is out of "regard to their own interest." Note that this does not at all preclude cooperation and help, or what came to be known as "altruism." It may well be genuinely felt, if felt at all. But it does preclude disinterested help given at personal cost.

Problems

Given his individualistic stand, Darwin at once faced two problems.[104] (I leave discussion of humans to the next section.) First what about hybrids? What about the mule, the offspring of the horse and the donkey? Notoriously, many hybrids—the mule being a paradigm— are sterile. But sterility obviously poses a problem for a theory like Darwin's, where the key is survival and even more (as Darwin realized fully) reproduction. Is there any point to sterility, meaning could selection have brought on sterility? This was certainly the opinion of Wallace who ever had a fondness for group selection. (As a lifelong socialist, Wallace had no truck with Adam Smith.) Explicitly challenging Darwin (in a series of letters in the 1860s), Wallace thought that hybrids are rather literally neither fish nor fowl. They cannot function as well as the members of their parent species. Hence for the benefit of the parent species, selection wipes out the inadequate mixed offspring. Better not to have them perpetuating, competing with the true offspring. Darwin could not see this. He could see that the offspring might not be as good as true offspring and therefore that selection could act to prevent organisms mating with members of other species. But once the offspring are born, then the parents have an interest in their survival and future reproduction. The interests of the individual count for all, and the interests of the species count for naught. Hence Darwin argued that sterility is just a nonadaptive by-product of the union of two disparate sexual systems. Mechanically things don't work, and selection has no role in the matter.

Somewhat more troublesome but in a way more intellectually

stimulating was the problem of sterility in the social insects. How is it that in groups like the ants and the bees and the wasps you find females who devote their whole lives to the well-being of others, their mothers and their siblings? Today, thanks to the work of the English evolutionist William Hamilton, this problem is cited as the paradigmatic example of individual selection in action, so-called kin selection.[105] Hamilton pointed out that when close relatives reproduce, in a sense one is reproducing by proxy because they are passing on copies of your units of heredity, the genes, and that in this day of genetics, passing on copies of your genes is what natural selection is all about.[106] He also noted that in the ants, the bees, and the wasps—the hymenoptera—because of a funny mating system, sisters are more closely related than mothers and daughters, so it pays to raise fertile sisters rather than fertile daughters. A worker's sterility pays dividends for that particular worker! Darwin of course had none of this, but he did spot that relatedness is the key. If in some sense your actions are helping the family, then selection can work on this.

> How the workers have been rendered sterile is a difficulty; but not much greater than that of any other striking modification of structure; for it can be shown that some insects and other articulate animals in a state of nature occasionally become sterile; and if such insects had been social, and it had been profitable to the community that a number should have been annually born capable of work, but incapable of procreation, I can see no very great difficulty in this being effected by natural selection.[107]

Darwin went on to point out that there is no great problem in explaining how it is that fertile animals might give rise to infertile ones. If the infertile are of value, then this can be carried on through the generations. It should be remembered that "selection may be applied to the family, as well as to the individual, and may thus gain the desired end. Thus, a well-flavoured vegetable is cooked, and the individual is destroyed; but the horticulturist sows seeds of the same stock, and confidently expects to get nearly the same variety; breeders of cattle wish the flesh and fat to be well marbled together; the animal has been slaughtered, but the breeder goes with confidence to the same family."[108] As always, the domestic case can now be flipped back analogically to nature. "Thus I believe it has been with social insects: a slight modification of structure, or instinct, correlated with the sterile condition of certain members of the community, has been advantageous to the community: consequently the fertile males and females of the same community flourished, and transmitted to their fertile offspring a tendency to produce sterile members having the same modification."[109]

Notice therefore that Darwin did allow selection to be applied to the family as well as the individual. The key is that families are interrelated groups, unlike most groups (including species). Was he inconsistent in this? Not really if you think back to the Adam Smith influence. Any sensible and successful businessman would have thought you were out of your mind if you suggested that he not pass on his wealth to his near kin. Josiah Wedgwood did precisely this in setting up his sons and daughters, including the parents of Charles and Emma. It was letting the money slip out of the family that worried people. Charles Darwin is the Platonic form of this kind of thinking. He married his first cousin, almost a business arrangement (although as it happens, a very happy marriage) that kept the family money intact. Then he lived at home catered to by an ever-concerned wife, surrounded by children. Trips away often involved going from one set of cousins to another. Even visits to London, to go to meetings or the dentist, usually involved staying with older brother Erasmus. I joke that the Darwin-Wedgwood clan could have given the Corleones lessons on family solidity. To Charles Darwin, the family was the individual, as the individual was the family.

Progress?

We come to the final issue to be discussed in this section. Although we are not now dealing directly with Darwin's views on humankind, humans are always there. And, at the end of the *Origin*, in the understatement of the century Darwin allows that his theory does have some interesting consequences for our species. Which raises the question of progress. For people like Erasmus Darwin, evolution was practically an excuse for progress, from the monad to the man as they said.[110] The same was true of later evolutionists, Lamarck and Chambers notably. Where did Charles Darwin stand on this issue? He believed in progress, with the system topped out by humans— preferably English-speaking humans. This was all part and parcel of the deist world picture he embraced. He may not have been with Lyell in thinking that humans required special miraculous interventions, but the picture would have been meaningless without our role in it. The deist-evolutionist-progressivist world picture was a genuine rival to the Anglican-theist-creationist-providentialist world picture, and the reason is that it too told a story of origins with humans as the climax.

However, from the start, Darwin knew he had to work at his progress. He never assumed that there is some kind of necessary force taking us up the chain of being to the top. For every business success, like the Wedgwood pottery works, there were a dozen or more failures. And what is success? Making money? Getting friends and influence? Enjoying what you are doing? Reflecting these tensions and ambiguities into his biology, we find that Darwin was never quite sure how you define the "top." "It is absurd to talk of one animal being higher than another.—We consider those, when the intellectual faculties [/] cerebral structure most developed, as highest.—A bee doubtless would when the instincts were—"[111] It really didn't seem that humans had an exclusive lien on the summit. "People often talk of the wonderful event of intellectual man appearing.—the appearance of insects with other senses is more wonderful; its mind more different probably, & introduction of man nothing compared to the first thinking being, although hard to draw line.—"[112] Above all, you have to work for success. It is not guaranteed. "The enormous *number* of animals in the world depends, of their varied structure & complexity.—hence as the forms became complicated, they opened fresh means of adding to their complexity.—but yet there is no necessary tendency in the simple animals to become complicated although all perhaps will have done so from the new relations caused by the advancing complexity of others."[113]

Darwin was always opposed to Germanic-type evolutionary progressions, based on the way that the embryo develops systematically from within to the full adult without the aid of outside forces. The philosopher Hegel was not an evolutionist, but he expressed the sentiment perfectly: "Nature is to be regarded as a *system of stages*, one arising necessarily from the other and being the proximate truth of the stage from which it results."[114] All going up to the climax, humankind. This was not Darwin's picture of things. "I grant there will generally be a tendency to advance in complexity of organisation," however, "the theory of natural selection . . . implies no necessary tendency to progression."[115] Moreover, the supposed analogy (made much of in some quarters) between the unidirectional development of the individual and the unidirectional development of the group is questionable.

> This general unity of type in great groups of organisms (including of course these morphological cases) displays itself in a most striking manner in the stages through which the fœtus passes. In early stage, the wing of bat, hoof, hand, paddle are not to be distinguished. At a still earlier [stage] there is no difference between fish, bird, &c. &c. and mammal. It is not that they cannot be distinguished, but the arteries [illegible]. It is not true that one passes through the form of a lower group, though no doubt fish more nearly related to fœtal state. (They pass through the same phases, but some, generally called the higher groups, are further metamorphosed.)[116]

He felt this way before discovering selection and even more as his thinking matured after the discovery. Necessity of this kind is simply alien to—denied by—evolution through selection. On the one hand, selection is relativistic. What succeeds in one situation may well not succeed in others. Being big is good if you are facing predators. No one takes on elephants. Being big is bad if you are facing famine. The mouse may succeed where the elephant does not. So there seems little steady climb to the top. Going off sideways or downward may be a good option. On the other hand, although Darwin knew little about the causes of variation, as noted he was convinced that they were random in the sense of not directed toward need. This was the source of tension with Gray who wanted precisely such direction. For Darwin, God's creativity comes exclusively in the force of selection, and that somehow obeys laws that are not especially guided.

Could there nevertheless be progress and could this be a selection-fueled progress? In the first edition of the *Origin*, Darwin certainly inclined toward progress as such. "The inhabitants of each successive period in the world's history have beaten their predecessors in the race for life, and are, in so far, higher in the scale of nature; and this may account for that vague yet ill-defined sentiment, felt by many palæontologists, that organisation on the whole has progressed."[117] He was also willing to give Louis Agassiz, non-evolutionist but firmly committed to individual-group developmental analogies, some slack. "I must follow Pictet and Huxley in thinking that the truth of this doctrine is very far from proved. Yet I fully expect to see it hereafter confirmed, at least in regard to subordinate groups, which have branched off from each other within comparatively recent times."[118] Although he added at once that this was all going to be a matter of selection. "I shall attempt to show that the adult differs from its embryo, owing to variations supervening at a not early age, and being inherited at a corresponding age. This process, whilst it leaves the embryo almost unaltered, continually adds, in the course of successive generations, more and more difference to the adult."[119] But finding a satisfactory definition of what it is that progresses and then linking it to selection had to wait until the third edition of the *Origin*. By now (admittedly only two years after the first edition) Darwin really felt he had got a handle on things. Progress is to be defined in terms of the physiological division of labor, namely, the differentiation of parts, and to get this Darwin invoked the idea of what we today call arms races, where lines of organisms compete against each other and improvement occurs, eventually, one trusts, the best improvement of them all.

If we look at the differentiation and specialisation of the several organs of each being when adult (and this will include the advancement of the brain for intellectual purposes) as the best standard of highness of organisation, natural selection clearly leads towards highness; for all physiologists admit that the specialisation of organs, inasmuch as they perform in this state their functions better, is an advantage to each being; and hence the accumulation of variations tending towards specialisation is within the scope of natural selection.[120]

Finally Darwin had an Adam Smith type of explanation to justify the sentiments expressed in the famous final paragraph of the *Origin*, one that dates back to an almost identical ending in his "Sketch" of 1842.

There is a simple grandeur in the view of life with its powers of growth, assimilation and reproduction, being originally breathed into matter under one or a few forms, and that whilst this our planet has gone circling on according to fixed laws, and land and water, in a cycle of change, have gone on replacing each other, and from so simple an origin, through the process of gradual selection of infinitesimal changes, endless forms most beautiful and most wonderful have been evolved.[121]

Even here the twin influences were at work, because (thanks to the overlap of words not to mention the sentiment) the passage obviously dates back to an already-quoted, Herschel-Whewell-influenced, I-want-to-be-the-Newton-of-biology paragraph, written in a private notebook in 1837, about the "powers" of forces working uninterrupted in biology as they do in astronomy. More than this though, it seems that Darwin then intensified the passage, echoing the natural theological sentiments to be found in a review of a book by Comte, written in 1838, by the Scottish man of science, David Brewster.

In considering our own globe as having its *origin* in a gaseous zone, thrown off by the rapidity of the solar rotation, and as consolidated by cooling from the chaos of its elements, we confirm rather than oppose the Mosaic cosmogony, whether allegorically or literally interpreted . . .

In the *grandeur* and universality of these *views*, we forget the insignificant beings which occupy and disturb the *planetary* domains. *Life* in all its *forms*, in all its restlessness, and in all its pageantry, disappears in the magnitude and remoteness of the perspective. The excited mind sees only the gorgeous fabric of the universe, recognises only its Divine architect, and ponders but on its *cycle* and desolation.[122]

Darwin had read this review when it appeared, with such intensity that it gave him a headache. As always, Darwin transformed what he read. Brewster was hymning God's interventionist, creative genius. Darwin was hymning Nature's noninterventionist, creative genius. As always, though, Darwin stayed firmly within his inherited culture-zone.

HUMANS

Much of the last two decades of Darwin's life was spent following ideas and facts that caught his fancy. Typical is the first book after the *Origin*, a little monograph on orchids, where Darwin showed again and again how he thought of organisms as machines, often exhibiting the very kinds of contrivances that were so vital to Britain's success in the nineteenth century.[123] This hearkens back to those early chemistry texts, as well as to his beloved botany teacher at Cambridge who urged his students to consider "all circumstances connected with the external configuration and internal structure of plants, which we here consider in much the same light as so many pieces of machinery," going on to "consider these machines as it were in action, to understand their mode of operation, and to appreciate the ends which each was intended to effect."[124]

Wallace's Apostasy

The big exception was the work on our own species.[125] In the *Origin*, Darwin stayed away from the topic because he did not want it to overwhelm the basic discussion, but (so he would not be accused of cowardice) making it clear at the end that we humans are at one with the rest of the living world. At once, however, others jumped right into the debate, and at once our species became the focal point of what came to be known as the "gorilla theory." Given his various social plans, it was inevitable that Thomas Henry Huxley would be a leader of this pack, first debating with Richard Owen about the nature of the human brain and then writing his own contribution.[126] Alfred Russel Wallace (figure 13), now back in England, was at first an enthusiastic coworker and he too took up the cause of humans, writing a paper much admired by Darwin where he argued that natural selection can be a major causal factor in the evolution of intelligence.[127] But then, like many other Victorians, Wallace got seduced by the allures of spiritualism, a conviction that lasted the rest of his life. Never one to let previous thinking or alliances influence his present enthusiasms, Wallace at once began arguing that there are many aspects of human nature that cannot be explained by natural selection.[128] At the physical level, he made mention of such things as human hairlessness. At the mental level, he argued that the full intelligence that we have — that is, that we Victorians have — cannot have been formed by selection because native people simply don't use that much intelligence. "Natural Selection could only have endowed the savage with a brain a little superior to that of an ape, whereas he actually possesses one but very little inferior to that of the average members of our learned societies."[129] Hence, spirit forces must have been involved.

Darwin was appalled, writing to Wallace: "I hope you have not murdered too completely your own and my child."[130] Something had to be done, and so Darwin was pulled into writing the *Descent*. For all that there are some very insightful moves, in respects as science it is not a groundbreaking work like the *Origin*; but, then, how many books are? It is more aimed at a general audience and often one senses that Darwin is coasting a little and relying heavily on the work of others. Yet is a work of huge significance for it bears out fully what I claimed in my prologue. We are in a new world. By now Darwin had moved from being a deist, as he was at the time of writing the *Origin*, to being an agnostic, and the *Descent* shows this. In the *Origin* there are about fifteen references to the "Creator," and many of them can be taken to imply that Darwin believed in such a being. (As in: "To my mind it accords better with what we know of the laws impressed on matter by the Creator, that the production and extinction of the past and present inhabitants of the world should have been due to secondary causes, like those determining the birth and death of the individual."[131]) In the *Descent* there are no such references that can be taken in that way. We are in the secular world. For all that the Catholic Church was about to deny its existence, Modernism had arrived. We are on the road, like it or not, to Richard Dawkins. Others as well as Darwin had directed us there—the advances in theology, the changes in social structure, and more. In fact, Darwin's own move to nonbelief was as much theological, an unwillingness to entertain the idea of the eternal damnation of nonbelievers (including his own father and brother), as anything. What is significant about the *Descent* is that it is so unstrained about it all. There is no special pleading. It is a fait accompli. It is now for us to work out the details and study the implications.

And yet. Modernism may have arrived, but its midwives had little idea about the infant they had ushered into the world. Taken as whole, the *Descent* is a book that testifies again and again to our main thesis—the work produced by Charles Darwin comes from and reflects the upper-middle-class society into which he was born, his family Whig legacy, and then his Anglican education. The man who wrote the *Descent* was not a man kicking against the pricks. Europeans are superior to other races. "When civilised nations come into contact with barbarians the struggle is short."[132] The British are superior to other Europeans. Quoting a classmate: "The careless, squalid, unaspiring Irishman multiplies like rabbits: the frugal, foreseeing, self-respecting, ambitious Scot, stern in his morality, spiritual in his faith, sagacious and disciplined in his intelligence, passes his best years in struggle and in celibacy, marries late, and leaves few behind him." Fortunately nature has something to say about this. "We have seen that the intemperate suffer from a high rate of mortality, and the extremely profligate leave few offspring."[133] In other words, the Irish have lots of kids but few survive whereas the Scots have but few but many survive. And then there is the case of women. Darwin's thinking could be taken straight from the pages of *David Copper-field*. Men are the big, strong, intellectual, hunter types. Women are the creatures of the heart—the angel in the house, as Coventry Patmore's appalling sentimental poem (1854–62) put it. "The chief distinction in the intellectual powers of the two sexes is shewn by man attaining to a higher eminence, in whatever he takes up, than woman can attain—whether requiring deep thought, reason, or imagination, or merely the use of the senses and hands." Embryological studies throw some light on this. "Male and female children resemble each other closely, like the young of so many other animals in which the adult sexes differ; they likewise resemble the mature female much more closely, than the mature male. The female, however, ultimately assumes certain distinctive characters, and in the formation of her skull, is said to be intermediate between the child and the man."[134] Not that we should be biased. There are good things to be said about the distaff side of the species. Indeed: "It is generally admitted that with woman the powers of intuition, of rapid perception, and perhaps of imitation, are more strongly marked than in man"; although not too much should be made of this: "some, at least, of these faculties are characteristic of the lower races, and therefore of a past and lower state of civilisation."

Topping things off, let us end with a few choice words on the virtues of capitalism.

In all civilised countries man accumulates property and bequeaths it to his children. So that the children in the same country do not by any means start fair in the race for success. But this is far from an un-mixed evil; for without the accumulation of capital the arts could not progress; and it is chiefly through their power that the civilised races have extended, and are now everywhere extending, their range, so as to take the place of the lower races. Nor does the moderate accumulation of wealth interfere with the process of selection. When a poor man becomes rich, his children enter trades or professions in which there is struggle enough, so that the able in body and mind succeed best. The presence of a body of well-instructed men, who have not to labour for their daily bread, is important to a degree which cannot be over-estimated; as all high intellectual work is carried on by them, and on such work material progress of all kinds mainly depends, not to mention other and higher advantages.[135]

Sexual Selection

A new species but expectedly (from one who began thinking in this way in the 1830s) the underlying themes are unchanged. Most importantly, for us as for all other organisms, it is evolution through selection that holds the causal key. What there is in the *Descent* is a shift of focus. Sexual selection had always been part of the picture, but now it was brought forward to an extent that the discussion was almost unbalanced. The reason was simple. Darwin agreed with Wallace that many of the features that make humans distinctive cannot be explained by natural selection. He thought, however, that sexual selection can do a lot of the heavy lifting. Things like hairlessness are the products of human tastes in beauty, and even intelligence can explained in this way. Alpha males get the choice of women, and so progress ensues. "The strongest and most vigorous men,—those who could best defend and hunt for their families, and during later times the chiefs or head-men,—those who were provided with the best weapons and who possessed the most property, such as a larger number of dogs or other animals, would have succeeded in rearing a greater average number of offspring, than would the weaker, poorer and lower members of the same tribes. There can, also, be no doubt that such men would generally have been able to select the more attractive women."[136] Obviously though they don't want fellow hunter-gatherers at home. They want sympathetic mates and mothers to their children. One species, two forms. "With respect to differences of this nature between man and woman, it is probable that sexual selection has played a very important part." Note that in an important way, this change of emphasis served to reinforce Darwin's strong commitment to the individual-organism approach to the workings of selection. Sexual selection is the paradigm of individual selection at work.

The focus on sex led Darwin to speculate about sex ratios, where he showed even more firmly his individual-selection stance. What are the facts, and what role does selection play in maintaining them? It all sounds a bit of a group problem—by definition sex ratios do involve the species—but Darwin's discussion made it clear that he never thought to solve the problem through group selection, suggesting that a species with one ratio will, because of that ratio, be naturally selected over other species. Rather, he tried to see how the group phenomenon comes about through individual selection. Most species have roughly equal numbers of male and female. Why is this so? Suppose to get a handle on the problem, a species is producing more males than females.

> Could the sexes be equalised through natural selection? We may feel sure, from all characters being variable, that certain pairs would produce a somewhat less excess of males over females than other pairs. The former, supposing the actual number of the offspring to remain constant, would necessarily produce more females, and would therefore be more productive. On the doctrine of chances a greater number of the offspring of the more productive pairs would survive; and these would inherit a tendency to procreate fewer males and more females. Thus a tendency towards the equalisation of the sexes would be brought about.[137]

Interestingly Darwin added that this equalization might come as a problem for the species, which now might be producing too many members to be sustained in their environment. He came up with another individual selection argument to explain how a species might get out of this jam, arguing (echoing the discussion about the Irish and the Scots) that although individual reduced fertility would be at a selective disadvantage, it might be accompanied by improved quality of offspring and so overall the less fertile might do better in the struggle. "By these steps, and by no others as far as I can see, natural selection under the above conditions of severe competition for food, would lead to the formation of a new race less fertile, but better adapted for survival, than the parent-race."[138]

Parenthetically it should be added that in the second edition of the *Descent* (1874) Darwin rather pulled back on all of this. He decided that he just couldn't see how individual selection could do the trick, even though ratios might be of benefit to the species "In no case, as far as we can see, would an inherited tendency to produce both sexes in equal numbers or to produce one sex in excess, be a direct advantage or disadvantage to certain individuals more than to others; for instance, an individual with a tendency to produce more males than females would not succeed better in the battle for life than an individual with an opposite tendency; and therefore a tendency of this kind could not be gained through natural selection."[139] He felt that he had been rather seduced by the group benefits and that was wrong. "I formerly thought that when a tendency to produce the two sexes in equal numbers was advantageous to the species, it would follow from natural selection, but I now see that the whole problem is so intricate that it is safer to leave its solution for the future."[140]

Religion

Recognizing therefore that Darwin had not lost his edge, let us turn now to the discussions in the *Descent* where, because Darwin was dealing with our species, he really did have to break new ground—religion and morality. First, religion. God was a big problem for many of the mid-Victorians. There were a number of contributing factors. Some felt that the so-called higher criticism showed that the Bible is far from divinely inspired but just another collection of ancient myths. Some thought (with Darwin) that many religious claims are more than false, they are pernicious. Some simply thought that, in the new urban, industrialized society, a religion like Christianity simply wasn't relevant. Darwin's treatment of the God question, significant if only because of its brevity, was very British and empirical. He calmly and simply adopted the position of David Hume, that religion really has no cognitive significance and is simply a by-product of other human characteristics. "We find human faces in the moon, armies in the clouds; and by a natural propensity, if not corrected by experience and reflection, ascribe malice or good-will to everything, that hurts or pleases us."[141] Darwin wrote of one of his dogs being disturbed by a parasol moving in the wind, and from this he concluded that religion is nothing but misinterpretations of naturally occurring phenomena. Going on the attack "every time that the parasol slightly moved, the dog growled fiercely and barked. He must, I think, have reasoned to himself in a rapid and unconscious manner, that movement without any apparent cause indicated the presence of some strange living agent, and that no stranger had a right to be on his territory."[142] It is hard to imagine anything farther from the tortured angst of continental thinkers faced with the collapse of traditional Christianity. Darwin was no Friedrich Nietzsche, railing at the death of God. Religion is all a big mistake.

Morality

Religion may not have been a big issue for Darwin. Morality certainly was. In the *Descent*, although he was sensitive to the issues, Darwin was not writing as a philosopher but as a mid-Victorian intellectual who was grappling with conduct after faith. Darwin would never have put things quite as dramatically or emotively as did George Eliot; few would, but her concerns were his concerns:

> I remember how, at Cambridge, I walked with her once in the Fellows' Garden of Trinity, on an evening of rainy May; and she, stirred somewhat beyond her wont, and taking as her text the three words which have been used so often as the inspiring trumpet-calls of men—the words *God, Immortality, Duty*—pronounced, with terrible earnestness, how inconceivable was the *first*, how unbelievable the *second*, and yet how peremptory and absolute the *third*. Never perhaps have sterner accents affirmed the sovereignty of impersonal and unrecompensing Law.[143]

Morality had to be discussed. In one sense, Darwin's thinking was fairly conventional. Talking about practice, no deep discussion was needed for the obvious answers—answers couched in terms acceptable to an upper-middle-class Englishman, one with liberal inclinations—so long as this includes a place for capitalism and opposition to unions ("a great evil for the future progress of mankind"[144]) and so forth. Where Darwin did start to show crucial and inventive insight was in his understanding that the underlying key to morality, particularly considered at the biological level, is some element of natural sociability. Most particularly, it has to be adaptive that we humans feel that we should get on with our fellows and that we have the ability to do so.

> The feeling of pleasure from society is probably an extension of the parental or filial affections, since the social instinct seems to be developed by the young remaining for a long time with their parents; and this extension may be attributed in part to habit, but chiefly to natural selection. With those animals which were benefited by living in close association, the individuals which took the greatest pleasure in society would best escape various dangers, whilst those that cared least for their comrades, and lived solitary, would perish in greater numbers. With respect to the origin of the parental and filial affections, which apparently lie at the base of the social instincts, we know not the steps by which they have been gained; but we may infer that it has been to a large extent through natural selection.[145]

Note the crucial role of "parental and filial" affections. This bears on what we shall have to say in a moment about Darwin's thinking on the individual–group selection issue.

Darwin realized that one had to go beyond this, and here he showed his background training and reading. As a young man he met and then read the works of James Mackintosh, Scottish lawyer and ethical philosopher, and had read Archdeacon Paley on moral philosophy when at Cambridge. So Darwin knew much about the basic moves needed for a full philosophical understanding of morality. In particular, Darwin was fully aware that morality is not just unreflective action, but at some level relies on conscious deliberation leading to action. In more modern language, one can say that Darwin appreciated that we have first-order desires, and then there are second-order reflections and thinking about these desires, leading to actions to tackle problems.

This is much bound up with the notion of conscience. Darwin had read Hume on animal reasoning and, from his earliest years, particularly from his earliest years as an evolutionist (that is to say the late 1830s), he saw a continuity between apes and humans. But he realized and stressed in the *Descent of Man* that it is the ability to reflect at second-order level that distinguishes moral beings, like humans, from mere unreflective animals. "The following proposition seems to me in a high degree probable—namely, that any animal whatever, endowed with well-marked social instincts, would inevitably acquire a moral sense or conscience, as soon as its intellectual powers had become as well, or nearly as well developed, as in man."[146]

There is still controversy among Darwin scholars about the extent to which Darwin was indebted to Hume directly. In the case of religion, we can trace influences with some confidence; the situation in morality is more complex.[147] Direct influence or otherwise, Darwin was with the British empiricists all of the way. For him, as for Hume, not to mention Edmund Burke and Adam Smith (whom Darwin flagged and whose work Darwin read as an undergraduate), the key notion was always that of "sympathy"—a kind of moral feeling or sentiment that one has for others. "The aid which we feel impelled to give to the helpless is mainly an incidental result of the instinct of sympathy, which was originally acquired as part of the social instincts, but subsequently rendered . . . more tender and more widely diffused."[148] Likewise: "Nor could we check our sympathy, even at the urging of hard reason, without deterioration in the noblest part of our nature. The surgeon may harden himself whilst performing an operation, for he knows that he is acting for the good of his patient."[149] Hume also used surgery as a place where sympathy has a rough time operating.

The Role of Natural Selection

What about the evolution of morality? How did Darwin think morality comes about? Natural selection is the causal focal point here. On the surface, the discussion was fairly straightforward. Those humans or proto-humans that were moral, that is to say who had a moral sentiment or sense of sympathy toward their fellow humans, were ahead of the game when it came to survival and reproduction. Morality therefore is a straightforward adaptation just like hands and eyes. The fact that morality involves thinking and behavior does not make it exceptional or peculiar. However, obviously, there has to be more to the story than this. How can morality evolve and persist given the possibility of cheating or of avoiding one's duty? Darwin himself put the matter bluntly:

> It is extremely doubtful whether the offspring of the more sympathetic and benevolent parents, or of those who were the most faithful to their comrades, would be reared in greater numbers than the children of selfish and treacherous parents of the same tribe. He who was ready to sacrifice his life, as many a savage has been, rather than betray his comrades, would often leave no offspring to inherit his noble nature. The bravest men, who were always willing to come to the front in war, and who freely risked their lives for others, would on an average perish in larger numbers than other men. Therefore it seems scarcely possible (bearing in mind that we are not here speaking of one tribe being victorious over another) that the number of men gifted with such virtues, or that the standard of their excellence, could be increased through natural selection, that is, by the survival of the fittest.[150]

Darwin grasped the nettle, giving this problem his fullest attention. For a start, he suggested that an important modifying factor might have been what is today known as "reciprocal altruism."[151] You scratch my back and I'll scratch yours. "In the first place, as the reasoning powers and foresight of the members became improved, each man would soon learn that if he aided his fellow-men, he would commonly receive aid in return. From this low motive he might acquire the habit of aiding his fellows; and the habit of performing benevolent actions certainly strengthens the feeling of sympathy which gives the first impulse to benevolent actions. Habits, moreover, followed during many generations probably tend to be inherited."[152] Then Darwin added: "But there is another and much more powerful stimulus to the development of the social virtues, namely, the praise and the blame of our fellow-men. The love of approbation and the dread of infamy, as well as the bestowal of praise or blame, are primarily due, as we have seen in the third chapter, to the instinct of sympathy; and this instinct no doubt was originally acquired, like all the other social instincts, through natural selection."[153] He elaborated: "To do good unto others—to do unto others as ye would they should do unto you,—is the foundation-stone of morality. It is, therefore, hardly possible to exaggerate the importance during rude times of the love of praise and the dread of blame." Adding: "A man who was not impelled by any deep, instinctive feeling, to sacrifice his life for the good of others, yet was roused to such actions by a sense of glory, would by his example excite the same wish for glory in other men, and would strengthen by exercise the noble feeling of admiration. He might thus do far more good to his tribe than by begetting offspring with a tendency to inherit his own high character."[154]

Note that none of this as such implies group selection. Praise and blame are rooted in sympathy and this, as we have seen, goes back to "parental and filial affections." At most we seem to have family selection. Obviously though this all bound up with culture—people passing the word around and expressing and formalizing thoughts and rules and expectations. So isn't it the case that ultimately he does agree that the tribe (figure 19) can be the unit of selection? He certainly does seem to imply this. "It must not be forgotten that although a high standard of morality gives but a slight or no advantage to each individual man and his children over the other men of the same tribe, yet that an advancement in the standard of morality and an increase in the number of well-endowed men will certainly give an immense advantage to one tribe over another." Darwin doesn't mince words about what this means: "There can be no doubt that a tribe including many members who, from possessing in a high degree the spirit of patriotism, fidelity, obedience, courage, and sympathy, were always ready to give aid to each other and to sacrifice themselves for the common good, would be victorious over most other tribes; and this would be natural selection." And so we get the consequence. "At all times throughout the world tribes have supplanted other tribes;

and as morality is one element in their success, the standard of morality and the number of well-endowed men will thus everywhere tend to rise and increase."[155]

Obviously there is a key question: What is a tribe? Darwin was not quite as explicit on this subject as one might have wished, a function probably of the fact that it was Wallace's prespiritualism paper on human evolution that inspired much of Darwin's thinking. In the context of human intelligence, Wallace wrote explicitly of selection between tribes. "Tribes in which such mental and moral qualities were predominant, would therefore have an advantage in the struggle for existence over other tribes in which they were less developed, would live and maintain their numbers, while the others would decrease and finally succumb."[156] The debt is obvious. The trouble is that for Wallace it would not matter at all if a tribe were simply a group of unrelated individuals. He was fine with this. But what of Darwin who clearly, whatever the case here, had individualistic tendencies? Did he just accept Wallace, or did he take over the language without truly making his own position clear? Does "tribe" mean "family"?

For Darwin, a tribe was certainly not a species. The *Descent* made it clear that selection works for the tribe against the species. "It is no argument against savage man being a social animal, that the tribes inhabiting adjacent districts are almost always at war with each other; for the social instincts never extend to all the individuals of the same species."[157] But are the members interrelated? Darwin suggested that tribe members think they are. He quoted Spencer on the origin of religion, where tribe members think that "names or nicknames given from some animal or other object to the early progenitors or founders of a tribe, are supposed after a long interval to represent the real progenitor of the tribe; and such animal or object is then naturally believed still to exist as a spirit, is held sacred, and worshipped as a god."[158] Then, when talking of the members of one tribe being absorbed into another, Darwin referred to the historian of ancient law Henry Maine and his claim that after a while these absorbed members believe that "they are the co-descendants of the same ancestors."[159] Most pertinently, talking of superior intelligence and of how it leads to the invention of useful artifacts, although Darwin followed Wallace in suggesting that this intelligence and its consequences could be a very significant adaptive advantage in the struggle to survive and reproduce, he linked the discussion with the point he made in the *Origin* about transmission of adaptations via close relatives.

> In a tribe thus rendered more numerous there would always be a rather better chance of the birth of other superior and inventive members. If such men left children to inherit their mental superiority, the chance of the birth of still more ingenious members would be somewhat better, and in a very small tribe decidedly better. Even if they left no children, the tribe would still include their blood-relations; and it has been ascertained by agriculturists that by preserving and breeding from the family of an animal, which when slaughtered was found to be valuable, the desired character has been obtained.[160]

The tribe is starting to look very much like an extended family—especially if you factor in the cultural point of what people think is the case may influence behavior even if it is not true biologically—and to gild the lily let us go to an article that the philosopher Henry Sidgwick wrote in 1876 in an opening issue of the journal *Mind*.[161] Darwin read it and was much intrigued by it, somewhat irritated perhaps. In an as-yet-unpublished letter to his son George, Darwin took up the matter, making it very clear that, as far as he was concerned, tribes are not groups of unrelated individuals. They are akin to the nests or groups that we find in the social insects.

To G. H. Darwin 27 April [1876]
Down Beckenham Kent
Ap. 27th

My dear George

I send "Mind"—it seems an excellent periodical—Sidgwicks article has interested me much.—It is wonderfully clear & makes me feel what a muddle-headed man I am.—I do not agree on one point, how-ever, with him. He speaks of moral men arising in a tribe, acciden-tally, i.e. by so-called spontaneous variation; but I have endeavoured to show that such men are created by love of glory, approbation &c &c.—However they appear the tribe as a tribe will be successful in the battle of life, like a hive of bees or nest of ants. We are off to London directly, but I am rather bad. Leonard comes home on May 10th !! Plans changed.[162]

There was reason why George would be the son to whom Darwin would direct this letter, for George was a brilliant mathematician and it was to him that Darwin had turned for advice when he was argu-ing with Wallace in the 1860s about the levels of selection problem.

To sum up. Morality is natural, it is essentially biological, and it is produced by natural selection. It is in other words an adaptation, one produced in the same way as the social instinct of bees and for much the same reason. Indeed, had things gone another way, the analogy would be even stronger. "If, for instance, to take an extreme case, men were reared under precisely the same conditions as hive-bees, there can hardly be a doubt that our unmarried females would, like the worker-bees, think it a sacred duty to kill their brothers, and mothers would strive to kill their fertile daughters; and no one would think of interfering." You could hardly have a more brutally natural-istic response to George Eliot's anguished cry. "The one course ought to have been followed, and the other ought not; the one would have been right and the other wrong."[163]

Darwin was a truly innovative thinker; he was not the Christian God. He did not make things out of nothing. Here as always Darwin took elements from the culture of his time and molded them to his own ends. Nothing new; everything new.

ENVOI

After the *Origin* was published the fact of evolution quickly became widely accepted, although of course there were those who were never convinced. We all know of the famous story of Thomas Henry Huxley clashing with the skeptical Bishop of Oxford.[164] In America there was and continued to be a large segment of the population who would never accept natural origins. At Harvard, the Swiss import Louis Agassiz went to his grave (in 1873) unconvinced of evolution.[165] At Princeton, the leading Presbyterian theologian Charles Hodge asked, "What is Darwinism?" And found his answer: "It is atheism."[166] But these were exceptions. Generally the scientific community (including Agassiz's students and even his son) came on board very quickly. A revealing item is that at Cambridge University, by the mid-1860s, on the final examinations in biology students were told to accept the truth of evolution and discuss the causes. Charles Darwin's son Frank got first-class honors, so someone was doing something right! And as went the scientific community, so went the general public. A little like the Hans Christian Andersen story of the emperor's new clothes, as soon as Charles Darwin—remember, already a well-known and respected and loved figure because of his wonderful story of the *Beagle* voyage—said "but we are monkeys," everyone said that they had known it all along.[167] Interestingly in the religious world also by about 1870 most people accepted evolution, and even evolution of the physical aspects of humankind. There were variations and exceptions. The Catholic Church became more and more conservative as the century drew on—primarily because of social and political events in Italy—and it turned against evolution.[168] But generally people were onside with evolution.

What about natural selection? Start with the general public reception. I argue Darwin's ideas came from his British culture. How did his British culture receive his ideas? With some understandable hesitations and qualifications—the conservative Prime Minister Benjamin Disraeli was not alone in worrying about the consequences for morality[169]—they were at once accepted right back into the system. To use a metaphor that would have been appreciated by one of Darwin's biggest partisans, Thomas Hardy, Darwin came from the soil of his homeland and his ideas returned to the soil of his homeland, where, watered by the creativity of the poets and novelists and essayists, they sprouted green shoots many times over. Natural selection became part of the landscape and in respects sexual selection even more so—no surprise really given that it focuses on just those issues that affect us all most intimately and emotionally.

So internalized was Darwinian thinking that people saw implications Darwin himself missed or was reluctant to draw. That fortunate grandson of the Industrial Revolution wanted to get some kind of progress out of the selection process, but his readers realized that for most of us any such progress was going to be very relative indeed. Take George Gissing's best-seller, *New Grub Street*, a novel that appeared ten years after Darwin's death. It is a tale of two writers: one talented but unwilling to compromise; the other far less talented but from the start fully aware of the struggle, of what it takes to win, and determined to be up there on the podium. There is natural selection as raw as anything to be found in the overheated stories of dogs written by the American novelist Jack London some years later.[170] Sexual selection is prominent too. "He was so human, and a youth of all but monastic seclusion had prepared her to love the man who aimed with frank energy at the joys of life. A taint of pedantry would have repelled her. She did not ask for high intellect or great attainments; but vivacity, courage, determination to succeed, were delightful to her senses." In the end, the loser dies, and the winner gets his wife too. She is happy with the arrangement, as well she might be: "though she had never opened one of Darwin's books, her knowledge of his main theories and illustrations was respectable."[171]

More subtle but as deeply Darwinian are the novels of Hardy. He saw that life is determined by chance events, and we do what we can with them, not always successfully. Some will win, but others will certainly lose. Poor Tess of the D'Urbervilles ends her life at the end of a rope, while her husband walks off with her sister. Hardy leaves unsaid what would be known to all of his readers, that that union was doomed to failure also. The law forbade a man from marrying his dead wife's sister. Fortunately for the mental state of the Victorians, not everyone took the Darwinian message in quite so somber a fashion, even when things went according to nature rather than according to plan. The poet Constance Naden tells of a young suitor who hoped that his studious nature would be irresistible. Alas, this was not to be.

But there comes an idealess lad,
    With a strut and a stare and a smirk;
And I watch, scientific, though sad,
    The Law of Selection at work.

Of Science he had not a trace,
    He seeks not the How and the Why,
But he sings with an amateur's grace,
    And he dances much better than I.

And we know the more dandified males
    By dance and by song win their wives—
'Tis a law that with *avis* prevails,
    And ever in *Homo* survives.

The poor sap is left with the reflection that there is nothing he could do. He was powerless against those never-ceasing laws.

Shall I rage as they whirl in the valse?
    Shall I sneer as they carol and coo?
Ah no! for since Chloe is false
    I'm certain that Darwin is true.

One could go on and on about Darwin's influence and how the mechanisms of natural and sexual selection became part of common culture—not always accepted, but respected and understood. Just one more example—a poem by W. B. Yeats, an Irishman to be sure but at a time when Ireland was part of the kingdom. He is well-known for his enthusiasm for theosophy and a host of other totally balmy esoteric systems. But he knew his Darwin, and in one of his more poignant poems this comes through starkly. A man has gone to consult an oracle, but all is drowned out by the cruelty of the struggle for existence in full flight.

O Rocky Voice,
Shall we in that great night rejoice?

What do we know but that we face
One another in this place?
But hush, for I have lost the theme,
Its joy or night seem but a dream;
Up there some hawk or owl has struck,
Dropping out of sky or rock,
A stricken rabbit is crying out,
And its cry distracts my thought.

If Yeats, of all people, can be gripped and moved by Darwinian processes in action, what need more to show that Darwin's thinking—
Darwin's mechanisms—became common currency?

This said, the story of natural selection in the scientific community is a little different. Everyone accepted some role for the process. It clearly could occur and had occurred. There was a group
that always thought natural selection was important and probably
the main cause of change. Not surprisingly, those who worked on
fast-breeding organisms with readily identifiable features, notably
butterflies and moths, led the way. Generally however selection was
relegated to a minor role.[172] There were a number of reasons for this,
starting with the fact that Darwin had no adequate theory of heredity
and people soon showed that without such a theory it was hard to
see how selection, no matter how powerful, could have any lasting
effect. The age of the earth was also a problem.[173] Ignorant of the
warming effects of radioactive decay, the physicists were allowing far
too little time for a leisurely process like natural selection. However,
the main determinant of the fate of natural selection was adaptation. There were some who did not think that selection unaided was
sufficiently powerful to account for adaptation. Darwin's American
supporter Asa Gray fell into this category, and he always wanted to
supplement selection with divinely guided variations.[174] More commonly however selection was ignored or relegated to second-class
status simply because people were not convinced that it was needed.
Selection is above all a force for explaining adaptation. If one down-
plays adaptation, then selection becomes an answer looking for a
question. By the 1850s, Owen and Huxley were with the tide in taking

the Germanic emphasis on homology as all-important. If one was going to do anatomy, if one was going to do embryology, increasingly if one was going to do paleontology, if one was doing systematics (as Darwin himself discovered when working on barnacles), then it was homology that one used as the daily tool of inquiry. Adaptations are if anything a nuisance, blocking out or misleading about true relationships. They point to analogies—shared functions and the like—not descent from shared ancestors. And from finding adaptations a nuisance, it was an easy step to finding them unimportant or even nonexistent. This is especially true if you are working with dead organisms—fossils or on the anatomist's dissecting table—where adaptations are totally nonfunctional.

With selection out of the picture, people were free to turn to other mechanisms, and much of the thinking owed huge debts to German Romantic philosophy and science. The emphasis on homology was all part and parcel of Romanticism's neo-Platonism, with all being seen ultimately as part of the One, the Absolute Unity. Embryology tied in with this, with unfurling of the organism being seen as a cameo of the unfurling of the evolving line. The belief, endorsed in a non-evolutionary fashion by Schelling's student Louis Agassiz, that in the stages of the individual's growth we see the stages of life's history, was now given a firmly evolutionary underpinning, and came up as the so-called biogenetic law: Ontogeny recapitulates phylogeny. After the *Origin*, the biggest booster of this neo-Romantic evolutionary world picture was the German biologist Ernst Haeckel.[175] He was, it is true, a deeply devoted disciple of Darwin, but do not be fooled by personal attachments or Darwin's friendliness toward him. Haeckel's world picture was not Darwin's world picture, nor was the kind of science he did the science of Darwin. This is not to say that Haeckel was not a good scientist. He was. It was just that he was not a Darwinian scientist. He, like many others, was much more interested in tracing out the paths of the evolutionary past than in focusing on the minutiae of causes. He was much more interested in seeing all as One—he was an ardent monist—than in sweating through the range of life-science areas that were covered by Darwin in the *Origin*. And above all, he was much more interested in making a quasi-religion with all bound together by thoughts of inevitable progress, culminating in humankind, than was Darwin. True, Darwin was a progressionist, but not a Romantic progressionist. Haeckel was.

Mine is a story about Darwin and not about Darwinism. I talk of the post-Darwinian period only to try to understand how an intelligent student of the subject could be so far off the mark as to think that Darwin's roots were not truly in his British heritage. Part of the reason has to be that most of us simply do not know how great was the effect of the Darwinian story on general culture as seen through poetry and fiction and the like. For those of us working on Darwin coming from a background in the history of science, disciplinary barriers have deterred us from seeing the huge amount of work and understanding now achieved by those in English literature studies and related fields.[176] But what happened in science also played its role, especially the "eclipse" of Darwin's thinking, as it was called by Thomas Henry Huxley's biologist grandson, Julian Huxley.[177] Much of the thinking about evolution in the fifty or more years after the publication of the *Origin*, by scientists writing as scientists,[178] owed little to Darwin's vision of the past and much to the Romantic vision, that of Goethe and Schelling and Oken and others. It was natural therefore to think that the real roots of the Darwinian Revolution lay in Germany and not in Britain. This was to start to change only around the 1930s when the great population geneticists—Ronald Fisher and J. B. S. Haldane in Britain and Sewall Wright in America—brought Darwinian selection and the new Mendelian genetics together into a synthesis ("neo-Darwinism"), thus providing the foundation on which modern evolutionary theory is based.[179] Although even then the change was fitful and the cultural debts of Darwin were often less (or made less) than obvious.

This in part was because the culture of Britain in the twentieth century was no longer the culture of Britain in the late eighteenth and early nineteenth centuries, even of the late nineteenth century. Or if it was, people felt uncomfortable and ashamed.[180] Some neo-Darwinians had horrendous social views about society, endorsing eugenics and thinking that unless the upper-middle classes are cherished all is doomed to degeneration and catastrophe. This was

Fisher's view, shared more recently by the greatest Darwinian of our generation, William Hamilton, who (as noted) explored and extended the implications of the individual-focused approach to selection, while backing this with beliefs that made Fisher look like a liberal.[181] He once told me that he regretted modern medicine because it meant the survival of so many inadequate members of society. Individual selection, popularized by Richard Dawkins as the "selfish gene" theory, seemed to be the harsh socioeconomic perspective of Margaret Thatcher translated into the language of biology. "There is no such thing as society. There are individual men and women, and there are families."[182] Natural selection applied to human beings, as it was by Harvard biologist Edward O. Wilson in his *Sociobiology: The New Synthesis*, was taken to be the distillation of sexist, capitalist, racist thinking in action.[183] Homophobic too. The natural desire therefore, as epitomized by the American cell biologist Lynn Margulis, who praised Darwin to the skies while at the same time condemning that "runt" neo-Darwinism, was to assume or pretend that Darwin himself belonged to a very different tradition.[184] Pertinently when people did stress the very Victorian views of Darwin, as, for instance, his views on unions—he shuddered at "the rule insisted on by all our Trades-Unions, that all workmen,—the good and bad, the strong and weak,—sh[oul]d all work for the same number of hours and receive the same wages"[185]—it was often done with the aim of undermining natural selection today or even evolution itself.[186]

This is one reason why there has often been a drive to separate Darwin from his real surroundings. This separation has also occurred in part because the Romantic vision was and still is powerful. This was so particularly of Sewall Wright who owed much to the thinking of Herbert Spencer who in turn owed much to the leading *Naturphilosoph* Friedrich Schelling.[187] Wright was ever a "holist," downplaying adaptation and natural selection, and endorsing group-enhancing mechanisms of change. This is a tradition that has continued, and particularly those who want to ameliorate what they take to be the harsh vision of pure Darwinism have been eager to see Darwin in their own terms. Thus, for instance, the Harvard paleontologist the late Stephen Jay Gould and the Harvard geneticist Richard Lewontin disliked the stress on adaptation, explicitly endorsing a more Germanic, homology-based approach, and were eager to underline Charles Darwin's ecumenical attitude toward evolution's mechanisms, where supposedly natural selection is but one among several possibilities. "We support Darwin's own pluralistic approach to identifying the agents of evolutionary change."[188] Likewise many in my own philosophical community are eager to divorce Darwin from his Whig family roots and make him a much friendlier figure when it comes to the harsh ways of natural selection. It is taken as a priori true that Darwin fell over himself in his eagerness to promote a group selection approach to evolutionary change. An example that is nigh egregious in its determination to make Darwin a figure that speaks to the present, paving the way to this kind of holism, is given in a recent book by the eminent philosopher Elliott Sober.[189] He is candid in telling "the story backwards to put my cards on the table at the outset," making it quite clear that his motivation in talking of Darwin at all is not for the sake of understanding Darwin but to find ammunition in his campaign for a warmer, less hostile world vision (he praises Lewontin) than is given by those who see struggle of one against all, mediated only by help given to others being returned as help given to self. Untroubled by a sensitivity to history, let alone a synoptic understanding of the pertinent creative thinking, this writer and his fellows make Darwin a sometime Young Hegelian who would have felt comfortable at family socials in Kentish Town, belting out duets around the kitchen table with his chum Friedrich Engels. The motivation by this side to make Darwin seem warmer and friendly by our criteria meshes happily with the motivation by the other side to make Darwin seem less cold and unfriendly by our criteria.

But as Thomas Kuhn implored us as historians of science, our job is to move into the past and not to judge it by our standards of today or to distort it for ends that we pursue. We must look at the past in its own terms, viewing it through the lens of the culture that was accepted then. Many of us probably feel very uncomfortable with the pure doctrines of Adam Smith and his fellows. We do not think workers are there simply to be exploited for our interests, and we think there are ways of controlling population growth other than workhouses of the most stringent and unpleasant kind. Likewise we are not enamored by Anglican Christianity, if we care for Christianity at all. I am certainly not saying that if we are to be Darwinians today, we have to subscribe to these past beliefs. No more am I saying that we should judge Darwin by our standards or (worse even) that we should not judge Darwin by his standards. On balance I think he was a really fine man, much loved by family, servants, friends and the public, although he did not always break from his upbringing in ways that he might and he was ruthless in using others for the ends of his science. (His leading biographer writes of "a sliver of ice" through the center of Charles Darwin.[190]) I am saying that if we are to understand Charles Darwin, yesterday or today, we must understand him in terms of his culture—the British culture described in the first section of this essay—whether we care for it or not. And we must not be seduced by cultures more congenial to us that were quite alien to this very great scientist.

FIGURE 1. William Paley (1743–1805), author of *Natural Theology; or, Evidences of the Existence and Attributes of the Deity* (1802). Painting by George Romney.

FIGURE 2. Adam Smith (1723–90). Artist unknown.

FIGURE 3. Thomas Robert Malthus (1766–1834).
Painting by John Linnell. Mezzotint.

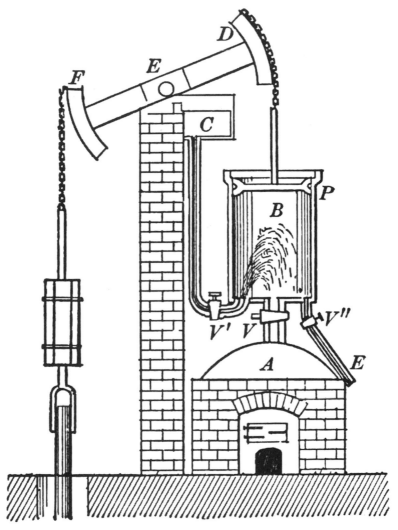

FIGURE 4. Newcomen engine, invented by Thomas Newcomen in 1712, most often used to pump water out of mines.

FIGURE 5. John Fredrick William Herschel (1792–1871).
Portrait by Henry Pickersgill.

FIGURE 6. William Whewell (1794–1866). Engraving by Eden Upton Eddis.

FIGURE 7. Jean-Baptiste de Lamarck (1774–1829).
Engraving by Ambroise Tardeu, 1821.

FIGURE 8. Erasmus Darwin (1731–1802), grandfather of Charles.
Portrait by Joseph Wright of Derby, 1770.

FIGURE 9. Charles Darwin (1809–82) at age seven.
From a chalk sketch by Ellen Sharples, 1816.

FIGURE 10. Alexander von Humboldt (1769–1859). Portrait by
F. C. Weitsch, painted shortly after Humboldt's return from
his five-year voyage (1799–1804) to the Americas.

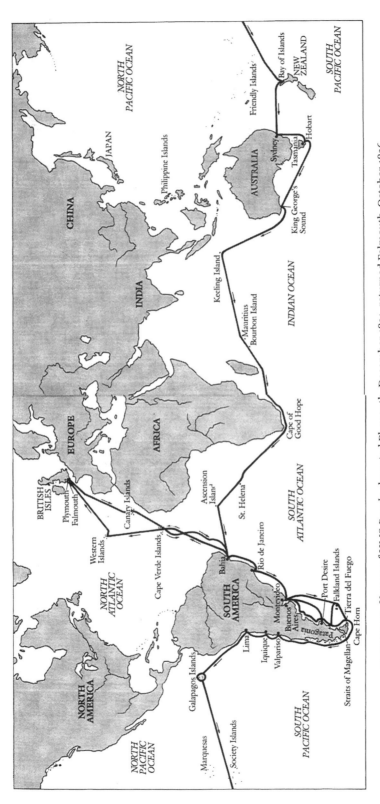

FIGURE 11. Voyage of HMS *Beagle*, departed Plymouth, December 1831, returned Falmouth, October 1836.

FIGURE 12. Charles Darwin, wedding portrait.
Chalk and watercolor by George Richmond, 1840.

FIGURE 13. Alfred Russel Wallace (1823–1913).
Oil over photograph by Thomas Sims, circa 1863–66.

FIGURE 14. Carl Gustav Carus (1789–1869). Portrait by
Julius Hübner, 1844. His arm is resting on his great anatomy book,
Von den Ur-Theilen des Knochen- und Schalengerüstes (1828).

FIGURE 15. Human and reptile skulls with the six bony plates
(transformed vertebrae) marked with Roman numerals. From Carl Gustav
Carus, *Von den Ur-Theilen des Knochen- und Schalengerüstes* (1828).

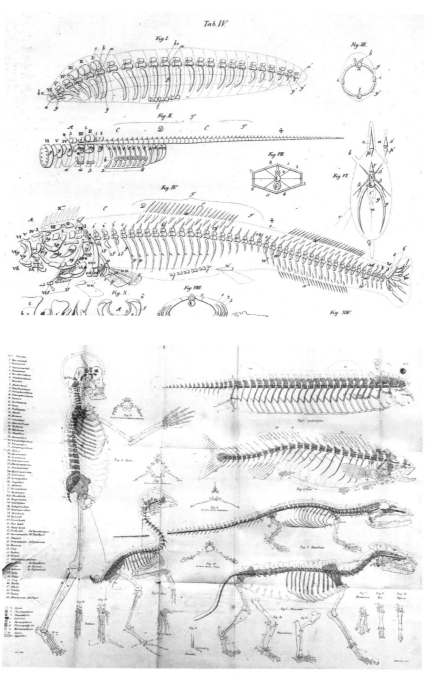

FIGURE 16. Above is Carus's illustration of a teleost fish, a jawless fish, and the Urtypus (or archetype), top; from *Von den Ur-Theilen des Knochen- und Schalengerüstes* (1828). Below are Richard Owen's comparable illustrations, with the archetype top right; from *The Nature of Limbs* (1849).

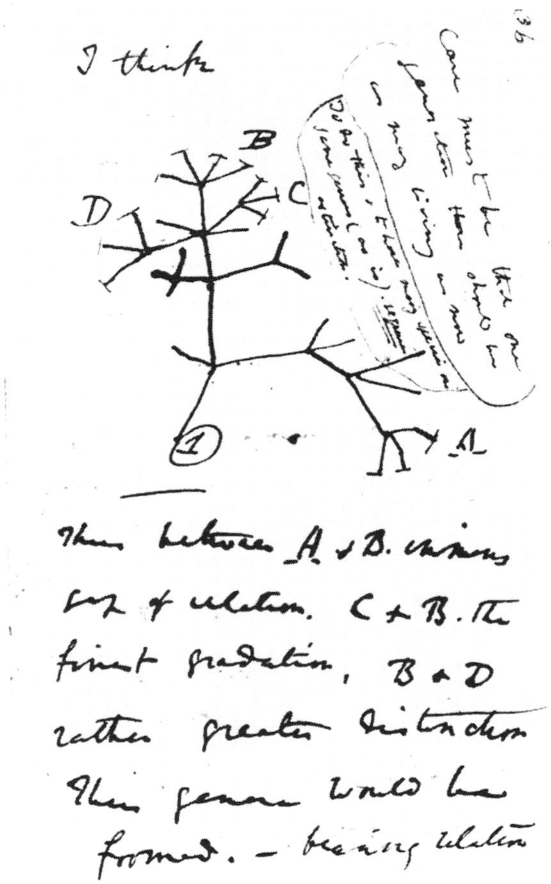

FIGURE 17. Descent tree (1837) from Darwin's *Notebook B*, depicting species branching with common ancestor and morphological forms represented at the nodes.

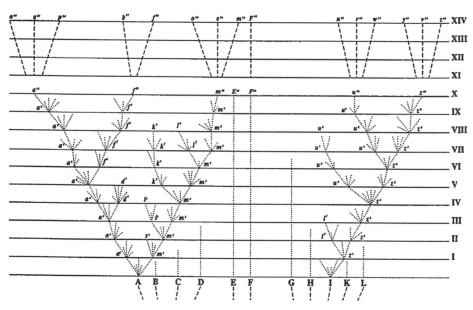

FIGURE 18. Diagram of species branching from Darwin's *On the Origin of Species* (1859).

ORANG UTAN ATTACKED BY DYAKS.

FIGURE 19. An attack by an "orang utan," an illustration
from *Alfred Russel Wallace's Malay Archipelago* (1869).

FIGURE 20. A spirit photograph of Wallace and an ectoplasmic
manifestation during a séance with Mrs. Guppy, March 14, 1874.
Courtesy of the College of Psychic Studies, London.

FIGURE 21. Charles Darwin, photograph by Elliot and Fry, 1881.

CHARLES DARWIN: COSMOPOLITAN THINKER

INTRODUCTION

In the preface to his play *Back to Methuselah* (1921), George Bernard Shaw expressed his century's growing fear of the dark specter of Darwinism:

> You cannot understand Moses without imagination nor Spurgeon [a famous preacher of the day] without metaphysics; but you can be a thorough going Neo-Darwinian without imagination, metaphysics, poetry, conscience, or decency. For "Natural Selection" has no moral significance: it deals with that part of evolution which has no purpose, no intelligence, and might more appropriately be called accidental selection, or better still, Un-natural Selection, since nothing is more unnatural than an accident. If it could be proved that the whole universe had been produced by such Selection, only rascals could bear to live.[1]

Many religious enthusiasts today share a comparable disdain for the apparent implications of Darwinian theory. They are not alone in their judgment. Most neo-Darwinians also agree with Shaw's assessment—something made explicit by such scientists and scholars as the late Stephen Jay Gould, as well as Richard Lewontin, E. O. Wilson, Jerry Coyne, Richard Dawkins, Daniel Dennett, and, to be sure, Michael Ruse. Despite Shaw's animadversions, these rascals seem to have borne up quite well confronted by an indifferent universe. But I would not count Darwin among them, at least not the Darwin who worked for two decades in the construction of an evolutionary theory, which he finally completed in 1859 with the publication of the *Origin of Species.* The Darwin of that book still heard the sweet birds sing in a universe responsive to human needs.

If we return to Darwin's accomplishment, now more attuned to the sounds of other, unexpected voices that guided him in the construction of his theory, we will come to appreciate a very different conception from the one that has seemed so familiar. I will argue that though Darwin heralded the modern age in biology, he was still a child of the early nineteenth century and formulated a theory that bore the century's time stamp. I think we have mistakenly assumed that Darwin banished final causes and notions of progress from biology, that he evacuated nature of moral purpose, and that he left a bare ruined choir, a universe Shaw found abhorrent. I rather believe that he deployed final causes to reconstruct nature with a moral spine and to provide it with a final goal, namely, human beings as moral agents. The resources upon which he relied in this understanding of nature derived from his own experience during the *Beagle* voyage and from the many observations and experiments he conducted after the voyage—but all of this as filtered through a conceptual matrix provided by the voluminous literature through which he came to perceive the activities of life. Especially prominent and determinative in this complex were treatises in natural theology and works stemming from the German Romantic movement of the late eighteenth and early nineteenth centuries.

Scholars have recognized the impact of natural theology on Darwin's thinking, though that impact is assumed to have been generally negative: natural theologians, like William Paley or William Whewell, provided the foil against which Darwin formulated his ideas of descent by natural selection. Gould, for instance, believed "the key to the logic of natural selection and its appeal for Darwin [lay] in the dual role of portraying every day and palpable events as the stuff of all evolution . . . and in overturning Paley's comfortable world by invoking the most radical of possible arguments."[2] Dennett likewise assumes that Darwin turned Paley's world upside down by postulating instead of intelligence behind organic design, rather a completely stupid "algorithmic process."[3] Ruse more astutely observes that Darwin's early natural theological convictions led him to regard an organism "as-if-it-were-designed-by-God."[4] But neither Gould nor Dennett nor even Ruse focus on a crucial distinction that structures Darwin's central argument in the *Origin*, namely, that while he opposed the notion of direct intervention of the deity in the production of organic adaptations, he nonetheless maintained that such adaptations were the result of divinely designed, intelligent law.[5] Throughout the period of the initial building up of his theory, he regarded natural law to be the product of a superior intelligence. As he affirmed to his American friend, the botanist Asa Gray, shortly after the publication of the *Origin*: "I am inclined to look at everything as resulting from designed laws, with the details, whether good or bad, left to the working out of what we may call chance."[6] Darwin perceived that natural law might absolve the Creator of responsibility for evil in the world, but he confessed to Gray his hesitating confusion about the issue. This confession signaled an acceleration of the long slide away from the faith he once held. By the late 1860s, he relinquished the notion of a superior intelligence formulating laws of evolution and described his attitude using the term newly minted by his friend T. H. Huxley: "agnostic."[7] Yet, and this is the point I wish to drive home, he constructed his theory initially under the assumption that mind was at work in the universe. This assumption brought with it a set of beliefs concerning the progressive development of creatures, the moral structure of nature, and the final goal of the evolutionary process. All of these beliefs were woven tightly into the fabric of the *Origin of Species*, and there is no evidence that Darwin ever abandoned them. They remained essential parts of the theory at his death.

A complementary strand of thought derived from the early German Romantic movement. Undoubtedly this claim seems simply implausible: that most sober of Englishman having his head turned by the wildly exotic ideas stemming from those individuals in the circle surrounding the poet Johann Wolfgang von Goethe! Yet, when one investigates Darwin's notebooks, reading lists, and essays, and surveys the names referred to—such names as Goethe himself, the Romantic scientist Alexander von Humboldt, and the anatomist and aesthetician Carl Gustav Carus—the suggestion may not seem so preposterous. And when the indirect routes to German idealism are explored—especially as charted by Johannes Müller and Ernst Haeckel, as well as by Richard Owen and William Whewell—the suggestion becomes more than an idle fancy, particularly in light of the distinctive ideas conveyed over these routes: the theory of the archetype, embryological recapitulation, and the notion that nature has a mental and moral structure.

If these two strands of thought—the natural theological and the German Romantic—have had a marked impact on Darwin's theory and endowed it with features that seem quite heterodox to modern biology, why hasn't this been observed before? Well, some few scholars have, indeed, urged these connections: Philip Sloan has highlighted Darwin's use of German sources;[8] John Brooke has explored the natural theological ideas that formed his early thinking about nature;[9] and John Cornell has emphasized Darwin's conceptions of teleology and designed law.[10] In the last few years, I also have begun to make the case.[11] But there are entrenched assumptions that have obscured these relationships and have fostered disbelief when the argument for them is made. At the outset here, I'll simply mention three such obfuscating attitudes.

First is the easy presumption that since Darwin's theory did ultimately give rise to our current neo-Darwinian beliefs, those beliefs must have characterized the original theory. Some of them, of course, do; but significant differences yet lie obscured in the haze of history. Abetting this presumption is a more technical, if informal, conception of the nature of theories. The logical empiricists have held theories to be abstract entities, types that might have different tokens. Bach's *Goldberg Variations* might be regarded as an abstract arrangement of tones that could be represented by Bach playing the variations on a harpsichord, Glenn Gould playing them on a modern piano, or a compact disk having those tonal relationships represented by a binary code, which itself is instantiated by pits in a polycarbonate layer of the disk. Comparably, one might regard Darwinian theory as a universal type that has instantiations in this or that representation in the contemporary period and in Darwin's own *Origin of Species*. The inference, then, is easy to make: our neo-Darwinian conception is essentially that of Darwin's own. This, I believe, is the casual assumption that has governed a large portion of scholarship on the Englishman's accomplishment. Nonetheless, I think it safe to say, Darwin was not a neo-Darwinian. A good deal of modern theory has simply been read back into Darwin's original conception.

A final source of the habit of reading modern ideas into the *Origin of Species* is our salutary effort to defend contemporary evolutionary theory against the onslaughts of religiously saturated claims to a non-Darwinian science—claims made by scientific creationists and intelligent designers. Since Darwin's theory (as conventionally understood) is the target of these religiously animated groups, there is an unreflective tendency to defend the progenitor theory of our modern biological science with all the resources of that science, and so amalgamate the latter with the former. In this defensive spirit, Darwinian science has been rearmed with instruments of materialism and atheism. In a quite typical fashion, Richard Lewontin, Steven Rose, and Leon Kamin so render the principal identifying construct of Darwin's science: "Natural selection theory and physiological reductionism were explosive and powerful enough statements of a research program to occasion the replacement of one ideology—of God—by another: a mechanical, materialist science."[12]

I will try to make good on my claims about the neglected but essential features of Darwin's theory in what follows. I would not deny that Darwin drew on many other sources: the importance of the work of Charles Lyell and Thomas Malthus have to be recognized, as most orthodox scholars do. The mistake, I believe, is to read these authors absent the context of Darwin's thought—that is, without consideration of his theological commitments and his appreciation of a network of ideas flowing out of Germany.[13] Initially I'll outline the major events in Darwin's life to provide a framework for the subsequent analyses of his works. I'll then begin to trace out the considerations I've mentioned by investigating the early notebooks, essays, and letters; I will also explore some of the twisting paths through his vast agenda of reading, reading that led to those little-recognized aspects of his conception of nature. Next, I'll discuss how these features, though often submerged, yet provided structure to Darwin's two principal works, the *Origin of Species* and the *Descent of Man*. Finally, I'll indicate what I think the implications of this historical analysis might be for our contemporary understanding of evolutionary theory—and for understanding ourselves, especially human consciousness and religious aspirations.

SKETCH OF DARWIN'S LIFE AND WORKS

Childhood through University

Charles Robert Darwin was born to Robert Waring Darwin (1766–1848) and Susannah (née Wedgwood, 1765–1817) Darwin on February 12, 1809, at the family home in Shrewsbury, England. Charles was the fifth of six children, including an elder brother and four sisters; the sisters served as surrogate mothers for their younger brother when Susannah Darwin died in 1817. Robert Darwin was a wealthy doctor and the son of Erasmus Darwin (1731–1802), who himself had advanced evolutionary proposals in *The Botanic Garden* (1789–91),[14] his effort at a poetical science that Coleridge compared to "mists that occasionally rise from the marshes at the foot of Parnassus."[15] The grandfather's (figure 8) more prosaic but influential treatise *Zoonomia; or The Laws of Life* (1796–98), which dealt with transmutation, heredity, and disease, won the admiration of the grandson and initially stimulated in him thoughts about the possibility of species change. Since *Zoonomia* was immediately translated into German and taken up by the Goethe circle, it made some of the circle's ideas preadapted, as it were, for the grandson's use.[16]

At age sixteen, Charles followed in the footsteps of grandfather, father, and elder brother Erasmus and enrolled at Edinburgh medical school. Darwin's father, however, seems to have held scant hope for his success. Darwin recalled in his *Autobiography* that his father once admonished him: "You care for nothing but shooting, dogs, and rat-catching and you will be a disgrace to yourself and all your family." Darwin immediately added to this recollection that his father, "who was the kindest man I ever knew," must have been out of sorts on this occasion.[17] At Edinburgh Darwin came into contact with Robert Grant (1793–1874), an expert on sponges and an advocate of Lamarck's transmutational hypothesis. He and Darwin became friends, and spoke about the evolutionary ideas of Lamarck and, likely, of Erasmus Darwin, whose books Grant knew quite well.[18] Darwin came down from Edinburgh after two years of a generally dreadful experience, though he did pick up, through Grant and his new friend and fellow student John Coldstream (1806–63), a taste for natural history.

Darwin felt the disappointment of his family, and so finally submitted to his father's insistence that he become a clergyman, a fitting profession for a younger son of the English gentry. After being crammed in Latin and Greek, Darwin enrolled (1828) at Christ College, Cambridge University. He felt no keen desire to pursue holy orders, but the idea of a country parsonage did have some appeal and, after all, he had to make good at something. Darwin spent a fair amount of his time at college collecting beetles and enjoying dinner parties, and thereby seemed to ratify his father's low expectations. But his time was not completely wasted. He became quite friendly with the botanist and polymath John Henslow (1796–1861), who introduced him to certain questions coming out of Germany about the origins of life and embryological development.[19] Through Henslow he also came to know several other of the dons, and occasionally walked home from Henslow's house with the formidable William Whewell (1794–1866). In order to pass out among those not seeking an honors degree, Darwin had to "get up" William Paley's (1743–1805) *Evidences of Christianity* and his *Moral Philosophy*. He read as well the theologian's *Natural Theology*, the logic of which gave him "as much delight as did Euclid."[20] But the book that inspired him as no other was *Personal Narrative of Travels to the Equinoctial Regions of the New Continent, during the Years 1799–1804* by the Romantic adventurer and friend of Goethe, Alexander von Humboldt (1769–1859).[21] Humboldt's (figure 10) aesthetic descriptions and exciting challenges kindled in the naive and parochial student a desire for exotic travel and research. He later avowed that "my whole course of life is due to having read & reread as a Youth [Humboldt's] Personal Narrative."[22]

The Beagle *Voyage*

Through the aid of Henslow, Darwin had opportunity to sail to the tropics on HMS *Beagle*. The *Beagle*, under the command of the twenty-seven-year-old Robert Fitzroy (1805–65), had the task of charting the waters off the coasts of South America, Australia, and the Pacific Islands. Darwin was to act as ship's naturalist and companion to the captain. The ship embarked from Plymouth harbor on December 27, 1831 and would not return until October 2, 1836. During the voyage Darwin and the manic-depressive Fitzroy had a rocky relationship. Darwin fancied Fitzroy could have been a Napoleon or a Nelson—he had that kind of commanding force; but he was also given to "austere silence," and indulged his "vanity & petulance."[23] As the ship lay at anchor in the various ports along the coast of South America, Darwin would travel inland to survey the geology (with Lyell's *Principles of Geology* as a guide) and to collect samples of the animal and plant life. He sent back many specimens to England, including the fossil remains of a giant Megatherium. The ship reached the Galapagos archipelago on September 15, 1835, and remained visiting the various islands of the group for about a month. Darwin noted many of the peculiarities of the fauna, especially the particular varieties of tortoise on the different islands. The only hint that he may have been thinking in terms congenial to his grandfather and Lamarck was a brief observation made on the return voyage. He recognized the similarity of fauna on the mainland of South America to that on the near islands, especially the animals on the Galapagos and Falklands: "If there is the slightest foundation for these remarks, the Zoology of Archipelagoes will be well worth examining; for such facts would undermine the stability of species."[24] Upon his return to London, after almost five years away from England, Darwin (figure 12) set to work cataloging his collections. In March of 1837, through conversations with John Gould, chief ornithologist of the British Museum, he became convinced that the three types of mockingbird he brought back from the Galapagos were not simply varieties of one species that had been altered by different environments, as he had originally supposed, but that they were good and true species. Thus what initially seemed merely varieties of the mainland species appeared to have breached the presumed species barrier. These reflections ignited a brain ready to explode with fresh ideas about natural history.[25]

Work on the Theory, 1837–59

While on the *Beagle*, Darwin kept a diary and notes on geology, zoology, birds, insects, and plants. The evidence from these sources indicates that he remained orthodox in biology for most of the journey, and only when sailing back to England did he jot in his ornithological notes the passage indicating some doubts about species stability, which I have cited above. In late May of 1836, while on the return, he opened his *Red Notebook* (so-called because of its cover), in which he made notes on geology and other matters; and he continued to post entries until spring of 1837. This notebook, in entries for March—recorded after the return—contains the first brief speculations on species change.[26] In July he began a geology notebook and a series of notebooks on the transmutation of species (his *Notebooks B, C, D, E, M,* and *N*); other scraps of notes reflecting on species have also survived from the period between 1837 and 1842.[27] These notebooks and loose compilations would furnish ideas for the first extended essay draft in which Darwin began to lay out his theory of species change by natural selection; this was his pencil sketch (thirty-five manuscript pages) of 1842. Two years later, he greatly expanded the essay to some 230 pages, for which he had a fair copy made lest he die before his theory could be published. These essays contain the skeleton of the *Origin of Species*.[28]

During the time Darwin worked on his species theory, he was also quite busily engaged in the publications that surveyed the results of his voyage. His diary and the geological and zoological notes kept on the *Beagle* supplied material for the first edition (1839) of his *Journal of Researches into the Geology and Natural History of the Various Countries visited by H.M.S. Beagle*. While certain passages in the book hinted at his new hypothesis about species, only in retrospect could they be so recognized. In the second edition (1845), several added passages alluded to the new perspective, but again these generally passed unnoticed. The zoology notebooks, as well as specimens sent back to London from the voyage, served as foundation for the five-part catalog of *The Zoology of H.M.S. Beagle* (1838–43), with introductions and supervision by Darwin but material described by various experts (e.g., Richard Owen, George Waterhouse, John Gould, etc.).[29] Darwin himself composed the three volumes of the *Geology of the Voyage of the Beagle* (1842–46).[30] In late 1846, he became interested in a small, quite unusual barnacle, one that lived within the shell of a mollusk. Intrigued with this new species, he began an exploration that would occupy him for the next eight years and would yield two folio volumes on living Cirripedia and two on fossil Cirripedia (1851–54).[31] This exhaustive study again hinted at his theory by implicitly endorsing the principle that ontogeny recapitulated phylogeny, a principle that appeared on the very initial page of his first transmutation notebook.[32]

After completing his work on barnacles, Darwin finally resolved to produce a big book on species that would spell out his theory and the complex evidence that supported it. His pocket diary records on May 14, 1856: "Began by Lyell's advice *writing* species sketch."[33] Lyell had urged Darwin to set down his theory in print, since there was the possibility that someone else might propose similar ideas. Darwin worked on his sketch into the following fall, and it had grown far beyond his initial intention. His expanding composition was to be called *Natural Selection*, and it would likely have gone to two fat volumes, crammed with evidence derived from his voluminous reading, his experiments on plants and in embryology, and his mathematical analyses of species patterns. However, the composition of the book was interrupted on June 18, 1858, by a letter from a sometime correspondent, Alfred Russel Wallace (figure 13), then in Borneo, where he was collecting specimens to be sent back to museums. He included in his letter an essay entitled "On the Tendency of Varieties to Depart Indefinitely from the Original Type."[34] Wallace wanted Darwin's opinion about the essay and requested that if it had merit would his correspondent send it along to Charles Lyell. Darwin was crushed. It seemed as if this obscure naturalist had been cribbing ideas from Darwin's own private essays, so similar were their conceptions. Darwin wrote Lyell: "Your words have come true with a vengeance that I sh'd be forestalled. . . . I never saw a more striking coincidence. If Wallace had my M.S. sketch written out in 1842 he could not have made a better short abstract."[35] Lyell and Hooker convinced their friend that honor did not require him to retire and give place to Wallace. These supporters arranged to have portions of Darwin's "Essay of 1844" (and excerpts of a letter to Asa Gray) and Wallace's essay read before the Linnaean Society on July 1, 1858. The event and the publication of their papers in the society's journal raised hardly an eyebrow. Darwin then set out to condense what he had already written for his big species book (some seven chapters) and quickly to add the remaining chapters that he had planned. His self-styled "abstract," *On the Origin of Species*, was published on November 24, 1859, in a run of 1,250 copies. It would eventually go through six editions; and with each, Darwin would add material and answer his critics. By the last edition of 1872 (and several subsequent printings) the book had been altered by about 50 percent.[36]

Darwin's *Origin* was a phenomenal success. In 1859, when it was published, one could hardly find a professional naturalist in Europe or America who believed in the transmutation of species. By 1882, when Darwin died, one could hardly find a professional naturalist who did not accept the evolution of species, even if many yet contended about the causes of species change. The force of the book's argument could not be denied. Nor could the man. Critics and colleagues alike succumbed to the character of this humble, genial Englishman, both as represented in his book and in his person. Ernst Haeckel, Darwin's great champion in Germany—and the individual most responsible for the warfare between evolutionary science and religion—visited Darwin at Down House in 1866 while on his way to the Canary Islands. He left this character impression of that first meeting:

As the coach pulled up to Darwin's ivy-covered country house, shaded by elms, out of the shadows of the vine-covered entrance came the great scientist himself to meet me. He had a tall, worthy form with the broad shoulders of Atlas, who carries a world of thought. He had a Jupiter-like forehead, high and broadly domed, similar to Goethe's and with deep furrows from the habit of mental work. His eyes were the friendliest and kindest, beshadowed by the roof of a protruding brow. His sensitive mouth was surrounded by a great silver-white full beard. The welcoming, warm expression of his whole face, the quiet and soft voice, the slow and thoughtful speech, the natural and open flow of ideas in conversation—all of this captured my whole heart during the first hours of our discussion. It was similar to the way his great book on first reading had earlier conquered my understanding by storm. I believed I had before me the kind of noble worldly wisdom of the Greek ancients, that of a Socrates or an Aristotle.[37]

Darwin's Other Projects

While continuously revising the *Origin* through subsequent editions, Darwin worked on an amazing number of other projects. In 1862 and 1865, he published on orchids and climbing plants; and in 1868 his two-volume *The Variation of Animals and Plants under Domestication* appeared.[38] In that latter work, he proposed a kind of genetic theory (his hypothesis of "pangenesis") that would accommodate the inheritance of acquired characters. His cousin, Francis Galton, attempted to test the hypothesis with transfusion experiments, which yield only negative results; Darwin, though, remained undaunted.[39] During the late 1860s, he began a series of exchanges on sexual selection with his new friend Wallace. As a result, he began to work on a book that would argue his particular version of sexual selection, which he thought a key to understanding the origin of the human races and sexual dimorphism in humans and animals. At the end of the decade, another dispute with Wallace broke out, a much more serious one this time. Wallace had converted to spiritualism and became convinced that higher spiritual powers were responsible for man's enlarged intellect and moral character. In the *Descent of Man and Selection in Relation to Sex* (1871), Darwin sought to give an extended analysis of sexual selection (in the second volume) and a detailed account of man's distinctive traits, especially his moral nature (in the first volume).[40] He thus responded to Wallace and others who doubted the capacious efficacy of natural selection. His theory of the evolution of morality would give rise to a very large response, positive and negative, from his time to ours—something I will consider toward the end of this essay.

Darwin had intended to discuss human and animal emotions at some length in the *Descent*, but with the book already projected to two volumes, he decided to publish separately his *Expression of the Emotions in Man and Animals* (1872). Two features of that book strike the modern reader. First, Darwin employed photographic illustrations, done by some London photographers and by the psychiatrist Duchenne de Boulogne, to show the similarity of emotional expressions among humans and to compare such expressions with those of animals.[41] Some of Duchenne's photos show how, for instance, the grimace of terror could be directly produced by galvanic stimulation of the facial muscles. The second curious aspect is that Darwin did not rely on natural selection to account for emotional expression; the inheritance of acquired habit carried the explanatory burden.[42]

During the last decade of his life, Darwin continued research on plants, publishing four more books, as well as a final dénouement on the lowly earthworm.[43] He died on April 19, 1882, without benefit of clergy, but nonetheless was buried with great ceremony in Westminster Abbey on April 25 of that year.

LITERATURE OF SIGNIFICANCE FOR DARWIN: ROMANTICISM AND NATURAL THEOLOGY

Though Darwin's experiences on his *Beagle* voyage were crucial to the development of his theory, many naturalists had spent extended periods in the lush jungles of South America and the islands of the South Pacific—and they did not produce an evolutionary theory. T. H. Huxley set out on a journey to South America, the Southern Pacific, and Australia that lasted from December 1846 to October 1850, almost as long as Darwin's own travel. During the trip, Huxley became an expert on various kinds of hydrozoa, and through the 1850s he published extensively on these wildly exotic invertebrates. Even with this intimate experience of mutagenic creatures and despite his rejection of creationism, Huxley did not become an evolutionist; rather, he opposed Lamarck's evolutionism and the comparable conceptions of his friend Herbert Spencer. Only when he read Darwin's *Origin* did Huxley become a convert. More was needed than raw experience and French theory.

Darwin's engagement with his grandfather's ideas and the introduction to Lamarck's theories that he received from Robert Grant undoubtedly primed him with the possibility of species change; and he had many idle hours during the *Beagle* voyage to connect such possibilities with the exotic creatures he had observed—his written notes, however, give only faint indication that his thoughts were drifting in the direction of the heterodox. Likely Lyell's *Principles of Geology* made its contribution, not only by convincing the young naturalist of the great age of the earth but also by sensitizing him to those slow processes that carved out large changes in the earth's surface.[44] Moreover, Lyell had undertaken an extensive review of Lamarck's theory in the second volume (1832) of *Principles*, a book that Darwin received by merchant ship while still in the early part of the voyage (fall of 1832). Later, when Spencer read both Lyell's account of Lamarck's theory and the geologist's subsequent rebuttal of that theory, he accepted the former and ignored the latter. Darwin certainly adopted the basic geological principles that Lyell had laid down; like Spencer, though, he apparently was not quite convinced by the antiprogressivism of Lyell's science. During the voyage the earth gave up to his shovel the fossils of giant creatures, primitive animals that no longer roamed the land. Lyell believed that eventually like creatures would reappear in a kind of steady-state, if oscillating, process, a process in which one kind of creature might go extinct while another kind, prefitted to its environment, would appear thanks to a divine hand keeping the balance of nature steady. Darwin seems not to have been convinced. The young naturalist's spare observation during the return voyage that similarity of creatures in the Falklands and Galapagos to those on the mainland might tend to undermine species— that remark suggests that, despite Lyell's animadversions, Lamarck's ideas, and those of his grandfather—were not far from his mind. So

Darwin's reading in evolutionary theory formed the matrix through which was filtered John Gould's assertion that the three mocking-bird types Darwin discovered in the Galapagos Islands constituted three good species. Darwin must at that point have recognized that the varieties of tortoise and the different finches that he saw might also be good species. His conventionally orthodox view that the environment could produce new varieties within the boundaries of a species seems to have been transformed, by the gentle shove from John Gould, into the unconventionally heterodox belief that many new varieties were in fact new species or, at least, incipient species. Well, this is a bit of speculation on my part, but quite historically reasonable I believe. Yet, there was an additional factor in the conceptual matrix that made Darwin's experiences during the *Beagle* voyage salient: the ideas of Alexander von Humboldt, whose *Personal Narrative* Darwin reckoned as one of the two most important books in his intellectual life (the other being Herschel's *Preliminary Discourse to the Study of Natural Philosophy*).[45]

In the remainder of this section, I will focus on the work of Humboldt and several other individuals whose proposals modulated Darwin's fecund mind, individuals who have not, I believe, been given their due weight in the formation of his theory. A consideration of their contributions will go far to show how German Romantic conceptions in particular got transmitted to English shores and were imbibed by Darwin. In subsequent sections, I will consider the other literature, Malthus's *Essay on the Principle of Population*, for example, that scholars have usually recognized as influencing the construction of his theory.

THE ROMANTIC FOUNDATIONS
OF DARWIN'S THEORY

Alexander von Humboldt

Alexander von Humboldt (1769–1859) and his brother Wilhelm (1767–1835) spent several years in Jena and became intimates of Johann Wolfgang von Goethe (1749–1832) and his circle, which in-

cluded Wilhelm (1767–1845) and Friedrich Schlegel (1772–1829)—the
founders of the early Romantic movement in Germany—the philosopher Friedrich Schelling (1775–1854), and the poet Friedrich Schiller
(1759–1805). Goethe and Alexander von Humboldt conducted anatomical research together in the mid-1790s. Later Goethe would say
of Humboldt that "he has a knowledge and vital wisdom whose like
we will not see again."[46] While in Jena, Humboldt prepared for an
extended research trip to South and Central America, something he
had been inspired to undertake because he had read Georg Forster's
(1754–94) celebrated account of Captain James Cook's (1728–79)
travels to the South Pacific.[47] Humboldt embarked in 1799 for the
new continent and spent some five years traveling in the northern
part of South America, along the Orinoco River, and up through
Mexico and finally sailing to the east coast of the United States. He
stopped in Philadelphia and Washington, where he spent six weeks
and had extended conversations with President Thomas Jefferson
(1743–1826). In summer of 1804, Humboldt and the companion of
his travels, Aimé Bonpland (1773–1858), returned to Europe.

Humboldt settled in Paris, where he became justly celebrated for
his research and especially for his harrowing climb of Chimborazo
on the Ecuadorian plateau—it was a volcanic mountain thought at
the time to be the highest in the world.[48] He had scaled the peak without, of course, any auxiliary oxygen; and bleeding from every orifice,
he just failed to reach the summit by a few hundred feet. The event
became Humboldt's signature accomplishment in the eyes of the
general public and garnered for him the sobriquet "Napoleón of Science," as well as "the second Columbus."[49] His fame spread to all of
Europe and America, even into the provincial corners of Cambridge,
England. For the next two decades in Paris, Humboldt employed his
notebooks to construct several monographs devoted to the various
branches of the sciences that concerned him—astronomy, geology,
plant systematics, and biogeography. He also composed a stirring
account of his travels in the multivolume *Personal Narrative*, and developed his aesthetic ideas in the essays gathered in the *Ansichten
der Natur* (Views of Nature, 1826). In 1827, Humboldt returned to
Berlin, where he lectured on his scientific endeavors to large pub-

lic audiences.[50] He also carried out several diplomatic missions to Paris for the Prussian Court. His lectures in Berlin became the basis for his five-volume *Kosmos*, the first volumes of which began to appear in 1845, when Humboldt was in his mid-seventies.[51] It was the last major work of his career. He died in that auspicious year of 1859.

There are four general features of Humboldt's conception that attracted the young Darwin: the sense of adventure and possibility in the study of nature; the notion that nature had to be understood as expressing systematically related laws; the awareness that aesthetic considerations offered a complementary approach to nature; and finally, that nature exhibited properties that had, in orthodox theology, been assigned to the deity.

While in his first year at Cambridge, Darwin had access to the initial volumes of Humboldt's *Personal Narrative*. By his senior year, he took to copying out long passages from the book and reading them to his friends, undoubtedly boring the more sporting of his circle.[52] The passages obviously had a magical effect on the young student, who began to cultivate the hope that he might, at least, visit the Canary Islands, which had been so evocatively described by Humboldt. Surely any young man whose horizons had been circumscribed by his schooling and position, who had failed at a profession prized by his family, but who had a vivid imagination and hopes for accomplishment—such a young man could hardly remain tethered to England's shores, at least in his reveries. Passages like the following must have turned the young Darwin's head to distant venues:

If America occupies no important place in the history of mankind, and of the ancient revolutions which have agitated the human race, it offers an ample field to the labours of the naturalist. On no other part of the Globe is he called upon more powerfully by nature, to raise himself to general ideas on the cause of the phenomena, and their natural connection. I shall not speak of that luxuriance of vegetation, that eternal spring of organic life, those climates varying by states as we climb the flanks of the cordilleras, and those majestic rivers which a celebrated writer [Chateaubriand] has described with so much graceful precision. The means which the new world affords the study of geology and natural philosophy in general are long since acknowledged. Happy the traveler who is conscious, that he has availed himself of the advantages of his position, and that he has added some new facts to the mass of those which were already acquired![53]

When Darwin set sail on the *Beagle*, he took along the first two volumes of the *Personal Narrative*, given to him as a parting gift by his mentor John Henslow, and he later acquired the remaining five. He also packed with him two of the German's geological works, and his political essay on Mexico. When docked in Rio de Janeiro he wrote his sister Catherine and requested her to have their brother Erasmus send along by merchant ship the *Tableaux de la nature*, the French translation of *Ansichten der Natur*.[54] While on the *Beagle*, these volumes prepared the young traveler for the exotic lands he would visit. On February 28, 1832, as the ship entered the Bay of All Saints, with the coastal city of Salvador at the far end and the area surrounded by the lush greenery of the jungle, an enraptured Darwin could only view the scene through Humboldtian eyes. He described it in his diary deploying the language of his predecessor:

> I believe from what I have seen Humboldts glorious descriptions are & will forever be unparalleled: but even he with his dark blue skies & the rare union of poetry with science which he so strongly displays when writing on tropical scenery, with all this falls far short of the truth. The delight on experiences in such times bewilders the mind. . . . The mind is a chaos of delight, out of which a world of future & more quiet pleasure will arise. —I am at present fit only to read Humboldt; he like another Sun illumines everything I behold.[55]

During the course of his five-year journey, Darwin would constantly invoke Humboldt's name, in letters home and in diary entries, when describing the passing of time in the tropics; the profusion and forms of vegetation; the constellations of the southern sky; volcanic formations; the color of landscape through the softening effects of the atmosphere; sickness in the tropics; the incursions of Christian missionaries; mountaineering; biogeographical relationships; and much more.[56] The language of Humboldt so weighted Darwin's descriptions that his sister Caroline, who loved his vivid accounts in the diary pages sent back to England, reproved him gently for using Humboldt's "phraseology . . . [and] the kind of flowery French expressions which he uses, instead of your own simply straight forward & the more agreeable style."[57] Darwin failed to accede to his sister's admonition. By comparing his later essays with comparable passages in the *Origin*, we can detect his efforts aesthetically to hone his language and to deploy the apt metaphor. He obviously kept in mind Humboldt's own prescription: namely, to provide not only an analytic account of natural phenomena but also to encourage the reader, through language and illustration, aesthetically to share the writer's experience of nature. Humboldt and Darwin both drew on the assumption that principally characterized the Romantic movement in science, namely, that aesthetic comprehension introduced and provided another avenue to the laws governing nature. Humboldt put it this way: as the naturalist explores the uncharted regions of the world, his mind everywhere "is penetrated by the same sense of the grandeur and vast expanse of nature, revealing to the soul, by a mysterious inspiration, the existence of laws that regulate the forces of the universe."[58] Humboldt recognized Darwin's efforts to be in harmony with his own. In his great work *Kosmos*, he attributed to the young author of the *Voyage of the Beagle* the same "aesthetic feelings" and "vividly fresh images" as found in the work of his own great friend, the adventurer Georg Forster.[59] When Darwin read that he was over the moon.[60] But the glow of admiration showed in both directions.

On the voyage back to England, Darwin had opportunity to consider his almost five years of travel around the globe and to assess the wealth of experience that had transformed him. He set out as a dilettante and naive student who had only a vague understanding of natural phenomena—mostly his lot of colorful beetles; he was returning as a man of scientific depth and tremendous feeling for the beauty and inner core of nature. He did not simply passively absorb the principles and aesthetic qualities of nature; he actively trained his eye to pierce the luxuriant flora to discover principles of relationship carried in the vistas of aesthetic delight. On the return voyage, while in those moments of reverie, he made explicit in his diary his debt to the German Romantic: "As the force of impression frequently depends on preconceived ideas, I may add that all mine were taken from the vivid descriptions in the Personal Narrative which far exceed in merit anything I have ever read on the subject."[61]

The nature that Darwin experienced through Humboldtian eyes did not clank along in the manner of a mechanical contrivance, nothing like a Big Ben of the Pampas. Rather Humboldt's nature and Darwin's as well bespoke intelligence, aesthetic depths, and even moral character. That nature was a cosmos, where land, climate, rivers, and mountains pulsated with the organic patterns of life. And through the exotic beauty of that cosmos, Darwin detected at the center the hegemonikon, the purposive principle of mind. He wrote in his diary:

> Among the scenes which are deeply impressed on my mind, none exceed in sublimity the primeval forests, undefaced by the hand of man, whether those of Brazil, where the powers of life are predominant, or those of Tierra del Fuego, where death & decay prevail. Both are temples filled with the varied productions of the God of Nature:—No one can stand unmoved in these solitudes, without feeling that there is more in man than the mere breath of his body.[62]

In the jungles of South America and the treacherous coasts around the Horn, Darwin found the God of nature, not the God of Abraham, Isaac, and Jacob.

Later, in the 1840s, Darwin examined the first two volumes of Humboldt's *Kosmos* in English translation. That book brought together in more systematic form the premise of much of Humboldt's research, namely, that events in the world were governed by general laws and that these laws mutually implicated one another in harmonious fashion. That symphony of laws made itself felt in the aesthetic features of nature experienced in the jungles of South America. These laws, from the abstract, analytic perspective, constituted the structural unity of the cosmos; from the particular, aesthetic perspective, they established the beauty of nature. Darwin, at the beginning of his theorizing and just before the publication of the *Origin of Species*, had thus absorbed from the multiple volumes of Humboldt his Romanticized conception of nature.

In later sections, I will explore Darwin's Romantic language, the specific tropes, metaphors, and personifications that bring nature vividly to life in the *Origin of Species*. Here I will consider Darwin's own reflections on the work of imagination in the production of scientific literature.

Darwin's "Castles in the Air"

Hugh Trevor-Roper observed that the great nineteenth-century historian Thomas Babington Macaulay confessed, in a letter to his sister, that his style was comparable to that animating the novels of Sir Walter Scott. Macaulay wrote: "My accuracy as to facts, I owe to a cause which many men would not confess. It is due to my love of castle-building. The past is in my mind soon constructed into a romance."[63] Macaulay's histories were indeed constrained by facts, but a vivid imagination plaited the strands tying the facts together, binding them now this way, now that, until a stable structure ensued, and one of aesthetic enchantment. Darwin as well believed that for the "discoverer" facts needed connections constructed by "building castles in the air." "Such castles in the air are," he urged to himself, "highly advantageous, before real train of inventive thoughts are brought into play."[64] The complementary joining of analytic faculties with imaginative capacities was exactly what Darwin found in the works of Humboldt, and what he would employ in fashioning the language of the *Origin*.

Taste for the imaginative realm undoubtedly led Darwin not only to Humboldt's great tales of adventure, which he consumed both before and after the voyage, but to speculate on more outré works of German Romanticism. What particularly caught his attention early in his speculations—just before reading Malthus—was a passage from Goethe's protégé Carl Gustav Carus (1789–1869) that had been recently translated into English. In Carus's "The Kingdoms of Nature, Their Life and Affinity," Darwin read of the unity of organic life

and mind, which connected diverse organisms throughout the kingdoms of animals and plants. He found the sentiments congenial and
adapted them to his new theory of descent, as he reflected:

> There is one living spirit, prevalent over this world, (subject to cer-
> tain contingencies of organic matter & chiefly heat), which assumes
> a multitude of forms each having acting principles according to sub-
> ordinate laws. There is one thinking . . . principle (intimately allied
> to one kind of organic matter—brain. . .). We see thus Unity in think-
> ing and acting principle in the various shades of separation between
> those individuals thus endowed, & community of mind, even in the
> tendency to delicate emotions between races, & recurrent habits in
> animals.[65]

The Romantic conception of higher powers coursing through nature
according to lawful determinations and furnishing the deep foundation of unity within the kingdoms of living creatures—that view
would form an intimate core of Darwin's developing understanding
of nature.

Like his master Goethe, Carl Gustav Carus (figure 14) was both
an anatomist and painter, really more accomplished in both than
his mentor, if not quite equipped with the same genius. He was also
a medical doctor, who, in 1813 as a young physician, became indel-
ibly acquainted with human anatomy as he worked in military hos-
pitals during the allied (Russia, Prussia, Austria, and Sweden) march
against Napoleon during the battle of Leipzig, the most costly and
ferocious of the Napoleonic Wars. It was a singular experience for the
young surgeon, since up to that time he had only theoretical knowledge from books about dealing with battle wounds.[66] That experience fueled his quest for greater anatomical and physiological understanding, a quest that was expressed in over a dozen subsequent
monographs. One of those volumes in particular had a considerable
influence on British natural history and ultimately on Darwin.

In his *Von den Ur-Theilen des Knochen- und Schalengerüstes* (1828),
Carus gave graphic currency to a conception that can be traced to
Kant and, more proximately, to Goethe, namely, that of an archetypal

idea, the pattern of which could be modified into various particular forms while retaining the primitive structure. Goethe first conceived of this in regard to plants. He contended that the various parts of a plant—the stem, leaves, calyx, petals, sexual organs—were modifications of a common, underlying structure, which he denominated the ideal leaf.[67] He adapted the idea to the vertebrate skeleton, suggesting that the various parts of the skeleton could be understood as modifications of a comparable underlying structure. Carus mentioned several other researchers in Germany and France that had the kernel of the idea; but it was Goethe's rival Lorenz Oken (1779–1851)—and Carus's friend—who really held the key. As Carus quoted him: "The entire system of bones is nothing other than a repeated vertebra."[68] Carus thus aimed to clarify and give definite form to the inchoate discovery of Goethe, Oken, and subsequent researchers. Just as Goethe had maintained the parts of the plant were transformations of a common, underlying structure—the ideal leaf—so Carus argued that the various parts of the vertebrate skeleton were transformations of an underlying structure, represented by the ideal vertebra. And he sought to trace these transformations up through ever more advanced vertebrate types and then back again. Goethe and Oken had both argued that the vertebrate skull ultimately consisted of six modified vertebrae; and Carus, following this conviction, focused his investigations on the skulls of a variety of vertebrate species. The detailed illustrations of his great book were unparalleled in their detail, and themselves made a compelling graphic argument. Beginning with a human being, Carus identified the six major bony plates that formed the skull (figure 15); he then traced back those morphologically similar plates to those of a reptile skull, then to the plates of a bird skull, finally to the structures of a teleost fish and to a jawless fish (figure 16). With each step back, the skulls came to look more and more like a series of vertebrae. Carus concluded this regressive analysis with the basic pattern of the vertebrate skeleton, which showed only the six vertebrae occupying the frontal position in a train of vertebrae (top illustration on left in figure 16). He found a grateful reader in the prominent British anatomist Richard Owen (1804–92), who modeled his own theory of the archetype on Carus's work and based his illustrations on those of his predecessor. Owen paid his debt to Carus (see figure 16) with only a vague attribution in his monograph *On the Archetype and Homologies of the Vertebrate Skeleton* (1848).[69]

Owen argued, retracing the discussions of Goethe, Oken, Geoffroy Saint Hilaire (1772–1844), and others whom Carus mentioned, that bones in vertebrates should be regarded as the same if they could be traced to the same bone or bones in the archetype. Such skeletal structures would express "general homology," which Owen defined in this way:

> A higher relation of homology is that in which a part or series of parts stand to the fundamental or general type, and its enunciation involves and implies a knowledge of the type on which a natural group of animals, the vertebrate for example, is constructed. Thus when the basilar process of the human occipital bone is determined to be the "centrum" or "body of the last cranial vertebra," its general homology is enunciated.[70]

The fundamental relationship of homology allows the naturalist to recognize that the bones of a bird's wing and those of the pectoral limb of a porpoise are archetypically the same, though only analogous to the structures in the wing of a dragonfly.

Shortly after it was published, Darwin read *On the Nature of Limbs*, where Owen repeated a version of his theory of the archetype and again used an illustration based on Carus's own. Carus worked out his conception of the archetype under the influence of the leading members of the Romantic movement in Germany: Goethe, Schelling, and Oken. They assumed that nature itself, under the Spinozistic formula of *deus sive natura* (God and nature were one), harbored developmental ideas. Owen, ever cautious about approaching too closely to the presumed atheistic idealism of the German Romantics, suggested that the archetype was an idea in the divine mind and that over time the Creator, employing the surrogate of nature, orchestrated the development of that idea into the various species of the vertebrate phylum.[71] Darwin, in responding to Owen's suggestion, returned the archetype to its Romantic source, namely, nature itself. He wrote on the back flyleaf of Owen's *On the Nature of Limbs*, "I looked at Owen's Archetypes as more than idea, as a real representation as far as the most consummate skill & loftiest generalization can represent the parent form of the Vertebrata."[72] For Darwin, the idea in God's mind had been transferred back to nature in the form of the primeval structure of the vertebrate ancestor: the multitude of vertebrate species had a common structure since they evolved from a common parent.

In the *Origin*, Darwin deployed the idea of the archetype in the penultimate chapter, in the section on morphology, where he reiterated his contention that archetypal structures—for example, the topological similarity of bones in the wing of a bat and the paddle of a porpoise—represented, not an idea in the mind of God, but a feature of the common ancestor. Significantly, Darwin drew support for his view by appeal to Goethe's theories of the ideal plant and the vertebral theory of the skull—thus aligning himself with these Romantic ideas.[73] Moreover, Darwin joined the idea of the archetype with another Romantic theory, namely, the recapitulation hypothesis—that the ontogenetic development of the embryo mirrored the phylogenetic development of the species.

Darwin complained after the publication of the *Origin* that his work in embryology was not given its due. In the *Origin*, the recapitulation hypothesis—that favorite of Romantic biologists, as Carus had made clear[74]—provided a strong support for his theory of descent. Darwin had endorsed the hypothesis on the very first page of his initial transmutation notebook, where he described the two kinds of generation his grandfather distinguished. Asexual generation occurred when a planarian split in two or plant cuttings produced new sprouts; but the "ordinary kind" was sexual generation, in which "the new individual passing through several stages (typical, . . . or shortened repetition of what the original molecule has done)."[75] That original molecule was the first "living filament," in Erasmus Darwin's view, that gave rise to "all the warm-blooded animals."[76] Subsequent individuals, according to the grandson, would thus pass through the same morphological stages as the original monad had passed through in its transmutational development.[77] In the *Origin*, Darwin put it this way: "Embryology rises greatly in interest, when we thus look at the embryo as a kind of picture, more or less obscured, of the common parent-form of each great class of animals."[78] The embryo, according to the hypothesis, thus resembles the adult parent form, and consequently recapitulates the morphological transitions of that ancient form over evolutionary time.

Darwin's theory of descent, as we have seen, owes a great debt, both directly and indirectly, to German Romantic biology. He followed the lead of that disciple of Goethe, Alexander von Humboldt, to embark on his trip of adventure at the beginning of his career. He endorsed several fundamental Romantic theses that formed the spine of his theory of descent: that of the archetype, of recapitulation, as well as the supportive vertebral theory of the skull and Goethe's conception of plant metamorphosis. Moreover, as I will argue in a later section, he, like Carus, rejected a mechanistic interpretation of nature; rather, he regarded nature as expressive of moral and aesthetic values, a nature directed to the production of human beings.

Overcoming Whewell's Kantianism

Sidney Smith, the editor of the *Edinburgh Review*, agreed with a friend who remarked of William Whewell (figure 6) that "science was his fort." Smith added, "and omniscience is his foible."[79] The master of Trinity College, Cambridge, held intellectual dominion over a vast array of subjects: a poet, who translated a fair amount of Goethe; a mathematician, who was a Second Wrangler at college; a professor of mineralogy; a theologian and philosopher influenced by Kant and the German idealists; and a historian of science, who composed a three-volume history of the inductive sciences that is still worth reading. While a student, Darwin had frequent occasion to walk home with him after both visited John Henslow, professor of botany and Darwin's friend and adviser. Darwin thought Whewell quite learned on "grave subjects."[80] About a year and a half after his return from the *Beagle* voyage, Darwin would renew his acquaintance with Whewell's formidable learning. In summer of 1838, he took up Whewell's newly published three-volume *History of the Inductive Sciences*. Therein he quickly perceived the challenge of Whewell's Kantianism.

Whewell developed a taste for deep, metaphysical philosophy while spending the summer of 1825 traveling through the Germanys, visiting university professors who specialized in mineralogy—he was preparing to vie for the open professorship of that subject at Cambridge. During the journey, his knowledge of German improved to such a degree that he later ventured to publish translations of Goethe's and Schiller's poetry. Under the influence of Coleridge and his disciples, Whewell also took up the volumes of Schelling and Hegel, hoping to make some headway in metaphysics. He began reading Kant's *Kritik der reinen Vernunft* during his tour or shortly thereafter, so his notebooks suggest.[81] Whewell's Kantianism brought him into a famous wrangle with John Stuart Mill over the status of mathematics and physics. Mill thought the basic principles of these disciplines derived from inductive conclusions drawn from experience, while Whewell argued that their universal and necessary character could not arise from particulars but had to originate through imposed ideas: a priori ideas were required, he thought, to bring experience to indisputable knowledge, the sort evinced by mathematics and physics. Whewell's Kantianism had another marked impact on his thought. Following the argument of Kant's *Kritik der Urteilskraft*, he refused to admit that there could be a scientific account of the origin and development of species.

In his *History of the Inductive Sciences*, Whewell demanded strict separation of science from theology, another Kantian stipulation. Science operated on the basis of empirical evidence and rational inference, which yielded explanatory laws, whereas theology depended on revelation and hope, which succored a faith in "things not seen." Science explored the law-governed framework of nature and the principles of its operation; theology unveiled the spiritual forces that erected the framework and authored its laws. From Whewell's perspective, one shared by such eminent biologists as Georges Cuvier and Louis Agassiz, biological organisms manifested teleological properties. In Cuvier's terms, they were subject to the "correlation of parts" and the "conditions of existence": internally, the parts of organisms were tightly knit together in means-ends relationships; and externally, the parts of organisms fit into their environmental stations with such precision that any significant alteration of parts — or the environment — would cause extinction of the organism and its type. Whewell agreed: clearly, organisms were purposively designed; indeed, our mind had been so molded in our interactions with the living world that we reflexively analyzed creatures employing concepts permeated with purpose. Nonetheless, as a Kantian, Whewell forbade a philosophical leap into the transcendent sphere to explain the designed structure of organisms. Science remained rooted in empirical observation and fixed law, and epistemological strictures confined its operations to the natural world.

The fossil evidence, according to Whewell, did indicate the extinction of ancient organisms and their replacement by progressively higher creatures. But this did not allow any inference of the sort made by Lamarck. Cuvier had shown that over long periods of time no fundamental alteration of species had occurred: mummies of humans, cats, and deer from Egyptian tombs remained recognizably the same as those living in Paris and in the woods around the city; moreover, the "conditions of existence" would have prevented fundamental species change. Both fact and theory thus argued that "species have a real existence in nature and a transmutation from one to another does not exist."[82] Since the scientist could not appeal to scripture for the needed miracles to explain the progressive replacement of species and since lawful physical causes did not avail, rational inquiry into the origin of species was forestalled. From a scientific point of view, the matter remained, as Whewell declared, "shrouded in mystery, and [was] not to be approached without reverence."[83]

Whewell thus set the problematic for the British biological disciplines in the first part of the nineteenth century: 1) organic life conformed to natural law, while also manifesting teleological, purposive structures; 2) fossil evidence indicated a progressive advance of species over time; 3) but no scientific argument employing natural laws could furnish an account of the intelligent design of organisms or their progressive appearance in the fossil record. The origin of species with their adaptive structures was thus scientifically intractable; only theology might advance a resolution of the mystery. (This is a position not unlike that of the scientific creationists and intelligent designers of today.)

Darwin recognized the conundrum for science that Whewell posed. In his "Essay of 1842," he remarked that Whewell closed off a scientific account of species, since that formidable thinker could not conceive of how organisms could become adapted to their surroundings while retaining species integrity.[84] It was Darwin's genius to show how theology could be naturalized, and thus provide a framework that both met Whewell's requirements for science and retained the archetypal intelligence within the realm of nature.

DARWIN'S SCIENTIFIC THEOLOGY

As a student at Cambridge, Darwin spent most of his time in the relaxed comfort of the gentleman, though as his B.A. exams approached, he again had to refurbish his classical languages and to contend with a bit of Euclid and some of the books of Paley. So the Cambridge years were not studded with great intellectual ferment; the orthodox attitudes of a seemingly conventional mind lay undisturbed.

The mature Darwin confessed astonishment at the religious convictions of his naive, younger self; he was simply abashed that as a student "he could believe in what he could not understand and what is in fact unintelligible."[85] Darwin wrote this last line in the mid-1860s, when he claimed for himself Huxley's neologism "agnostic." But in his *Autobiography*, he yet acknowledged something obvious from his notebooks, essays, and the *Origin* itself. He revealed that at the time he had completed the *Origin*, though he had lost his faith in Christianity and its doctrine of hellfire, he still retained a belief in a "First Cause having an intelligent mind in some degree analogous to that of man."[86] What should we make of this? Did his belief in an intelligent mind governing the universe have any impact on his science?

In his 1842 essay, Darwin asserted that natural selection was a law that explained the development of species. He yet recognized that natural laws, especially those that could give an account of complex adaptations, themselves need an explanation. Using quite traditional language, he called these laws "secondary causes." He put it this way at the end of the essay, in a passage that begins to formulate the eloquent conclusion of the *Origin*:

> There is much grandeur in looking at the existing animals either as the lineal descendants of the forms buried under thousand feet of mater, or as the coheirs of some still more ancient ancestor. It accords with what we know of the law impressed on mater by the Creator, that the creation and extinction of forms, like the birth and death of individuals should be the effect of secondary [laws] means. It is derogatory that the Creator of countless systems of worlds should have created each of the myriads of creeping parasites and [slimy] worms.[87]

Darwin thus insisted, early in the formulation of his theory and in the *Origin*, that the laws he discriminated could explain the origin of species, but that a primary cause, an intelligent mind, was needed to explain the laws themselves. Darwin made this clear to his readers by the epigrams he chose to face the title page of the *Origin*, one from Francis Bacon and one from William Whewell, both carrying the same message. Darwin quoted from Whewell's *Bridgewater Treatise*:

> But with regard to the material world, we can at least go so far as this—we can perceive that events are brought about not by insulated interpositions of Divine power, exerted in each particular case, but by the establishment of general laws.[88]

Several times in the *Origin*, Darwin made an appeal to the distinction between secondary laws or causes, of which natural selection was the most important, and the primary cause, namely, the Creator.[89] So for example: "To my mind it accords better with what we know of the laws impressed on matter by the Creator, that the production and extinction of the past and present inhabitants of the world should have been due to secondary causes."[90] Some months after the publication of the *Origin*, several antagonistic readers—Richard Owen and Adam Sedgwick, for example—ignored Darwin's explicit inclusion of God as the ultimate cause of the structure of nature, if not of the fine details of its operation. He complained to his friend Asa Gray that he "did not intended to write atheistically." While he did admit he was confused about the issue, he nonetheless confessed, in a line that echoed one from Herschel: "I am inclined to look at everything as resulting from designed laws, with the details, whether good or bad, left to the working out of what we may call chance."[91] During the 1860s, Darwin continued down the long slide away from religion, and eventually gave up the idea of an intelligent mind governing the universe; or rather, he thought the matter too mysterious to be rationally comprehended. He simply adopted Huxley's term "agnostic" to characterize his religious views.[92]

The significant point to stress, however, is that the long argument of the *Origin of Species* was constructed under the assumption of a governing mind, and that assumption had consequences. Ruse, though, seems convinced that the belief in a governing mind had no scientific relevance for Darwin's theory; it was simply a matter of his theological opinion, obiter dicta without standing. Ruse puts it this way:

> Darwin proceeded by breaking down the argument to design into a scientific part and a nonscientific part, giving an answer (natural selection) to the scientific part, and then saying that the nonscientific part is really not his concern as a scientist.[93]

Despite the *Origin*'s explicit appeal to a primary cause to explain the secondary laws governing the actions of nature, Ruse concludes that the assumption simply played no role in Darwin's science. Ruse is not alone in this conviction. Elliott Sober also attempts to exculpate Darwin's theory of supernatural taint by claiming that the English master's explanatory appeal to God as a first cause was a philosophical "argument for the existence of God," rather than any scientific feature of his theory. Sober maintains that the notion of God as primary cause didn't penetrate or shape Darwin's science.[94]

Both Ruse and Sober suppose that a nice distinction between science and philosophy existed during the period of Darwin's composition of the *Origin of Species* (1837–59). But there is no such fast distinction until the end of the nineteenth century. From the early modern period through the nineteenth century, "natural philosophy" would be a common rubric for a systematic approach to such topics as the status of species, their origin, and their development. Toward the end of the century, the term "science" would come ever closer to what we mean by the word, but through mid-century one would find "philosophy" and "science" used interchangeably—experimental philosophy or experimental science, moral philosophy or moral science.[95] Even Whewell, who did the most to promote "science" as distinctive designation—and coined the term "scientist"—would use, for instance, "physical philosophy" interchangeably with natural science.[96] Michael Faraday referred to himself as an "experimental philosopher" to the end of his days.[97] Darwin provides a pertinent example of this interchangeability when in the *Origin* he introduced the concept of *struggle for existence*: "The elder De Candolle and Lyell have largely and *philosophically* shown that all organic beings are exposed to severe competition."[98] So it is perfectly arbitrary for Ruse to exclude the explanation for natural law from Darwin's theory on the assumption that such explanation falls into the waste bucket of philosophy. Darwin was perfectly explicit about both the explanation of natural law as a secondary cause and the very meaning of the fundamental concept of his theory: "By nature, I mean the laws ordained by God to govern the Universe."[99] And Sober?—his effort to impose the distinction between science and philosophy on Darwin's theory has severed even tenuous threads to plausibility: there is simply no evidence that Darwin was trying to prove the existence of God; he, rather, assumed it. Darwin was a nineteenth-century thinker, and all evidence shows that with complete sincerity he asserted a proposition that would not have been exceptional in his time: a primary cause established natural laws and explained their character. I will specify below more exactly what the consequences were of Darwin's proposition that God was the primary cause of the laws of nature.

Ideas flowing from both the German Romantic tradition and from more orthodox natural philosophy gave essential form to Darwin's theory. Even good historians have been blinded by the light of modern evolutionary theory when attempting to give an account of the historical Darwin. In that brilliant glow coming from our contemporary science, those historians have constructed his doppelgänger, a creature who has become ubiquitous in historical introductions to biology textbooks and in the more casual approaches to the *Origin of Species*.

DARWIN'S CONSTRUCTION OF HIS THEORY

Explanation of Transmutation

Darwin was, as the Freudians might put it, anal-retentive. He stored away his various notebooks, reading notes, essays, drafts, and letters—now readily available to scholars—so that we are able to follow him in the construction of his theory in satisfying detail, descending to greater depths than possible with almost any other major scientist. In what follows I will pursue the quarry, though focusing on the more significant features of Darwin's effort.

After he returned from his *Beagle* voyage in October of 1836, Darwin began immediately working through the great hoard of biological and geological specimens he had sent back to the British Museum. With the help of individuals like John Gould, the museum's principal ornithologist, and Richard Owen, the Hunterian Professor of Anatomy, Darwin and his coworkers began systematic descriptions of those specimens. As mentioned above, it was Gould who pushed Darwin over the edge about species stability. Gould determined that the different types of mockingbird Darwin brought back from the Galapagos were not mere varieties of the mainland species, as the young adventurer thought, but that they were good species. This obviously tripped a mind at the ready.

Though Darwin was quite familiar with the transmutational theories of Lamarck and his own grandfather, he initially was unclear about the causes of alterations in species. A theory he considered early on seems to have derived from reading Lyell's brief account of the views of the Italian geologist and paleontologist Giambattista Brocchi (1772–1826), who held that species were like individuals: they were born at a certain time, lived for a period, and died, that is, went extinct. In a notebook entry of March 1837, one that recorded evidence of what seemed to be the abrupt replacement of one species by a different but similar species, Darwin jotted: "Tempt to believe animals created for a definite time:—not extinguished by change of circumstances." This idea lingered, and found its way into Darwin's *Researches* (1839) of the voyage of the *Beagle*: "All that at present can be said with certainty is that, as with the individual, so with the species, the hour of life has run its course, and is spent."[100] A few months after he ventured the senescence theory of species extinction, he weighed it against both the Lamarckian proposal that species changed constantly under direct environmental impact, and Lyell's supposition that when the environment changed too rapidly, a species would go extinct. By summer of 1837, Darwin became convinced that "death of species is a consequence . . . of non adaptation of circumstances."[101]

Concomitantly with the senescence theory of extinction, Darwin supposed that the environment might directly alter individuals of a species, molding them to its requirements. This Lamarckian-like supposition became part of the permanent deposit of Darwin's theory. Yet the direct effects of the environment as a means of species change seemed to him a bit crude. He subsequently introduced a more flexible device, namely, the inheritance of habit, something Lamarck had also proposed; Darwin, though, seems to have found the clue in the writings of Frédéric Cuvier (1773–1838), Georges Cuvier's younger brother. Darwin considered, for example, how exercise might have modified the foot of the jaguar: "Fish being excessively abundant & tempting the Jaguar to use its feet much in swimming, & every development giving greater vigour to the parent tending to produce effect on offspring—but whole race of that species must take to that particular habitat.—All structures either direct effect of habit, or hereditary & combined effect of habit."[102] So prior to reading Malthus, Darwin had two causal devices by which species might be altered: the direct impact of the environment and the inherited effects of habit. But he seems not to have been quite satisfied with these devices, and continued to think about the problem.

Reading Malthus's essay on population would provide the framework for a more satisfactory account of species change. But many individuals read Malthus and did not hit upon natural selection. Only a mind prepared might flash into invention. In his *Autobiography*, Darwin mentioned a salient consideration that led him to perceive the significance of Malthus—and it was not any Newtonian mechanistic analogy, though Ruse thinks otherwise.[103] He came to appreciate the power of what we call artificial selection. But there is, I think, another idea that also primed him when reading Malthus; his notebooks reveal that he was thinking long and hard about *the struggle for existence*. Let me take these two considerations in turn.

Darwin recognized that domestic species had their structures changed as a result of the breeders' art; and he was certainly aware that his grandfather and Lamarck (figure 7) had employed the model of domestic breeding as a clue for the operations of nature. Despite Lyell's objection that domestic animals had been especially chosen by primitive man because of their more plastic character—thus artificial breeding could hardly serve as a model for natural processes[104]—Darwin, in summer of 1838, began speaking with breeders and reading the manuals of practitioners like Sir John Sebright and John Wilkinson.[105] This literature brought him to understand the power of domestic "selection" (Sebright's term), as Ruse has shown.[106] Yet Darwin was initially puzzled, as to what might play the role of the natural selector or "picker." Quite obviously a human intelligence operated in the domestic situation, both to pick or select individuals because of certain traits and to separate those individuals from less endowed organisms and to breed only from the former. Yet what was the picking agent in nature? Darwin recognized the problem in midsummer of 1838: "in a really natural breed, not one is picked out."[107] How could selection occur in nature when no intelligent agent was picking the few "best individuals," segregating them from the others, and breeding from just these few? Darwin would shortly find that intelligent agent.

The other important idea that Darwin formulated before reading Malthus was inspired by a line from Lyell's *Principles of Geology*. The geologist had cited de Candolle's observation that all of the plants of a country "are at war with one another."[108] Perhaps it was the image of plants taking up arms, troops of daffodils, brigades of bamboo, snapdragon sappers on the march—certainly an arresting analogy. Lyell believed the struggle of organisms accounted both for the stability of species and, when the struggle became overwhelming, for the extinction of species. Darwin took to heart Lyell's implied admonition to "study the wars of organic being," and recounted in his notebook the victories of guavas in Tahiti and thistles in the Pampas.[109]

When Darwin took up Malthus's *Essay on the Principle of Population* for "amusement," in late September of 1838, he detected in the volume what others (perhaps save Wallace) did not. He discovered there a way of thinking about reproduction and its effects. Malthus (figure 3) argued that a vigorous and growing human population would eventually outstrip its food supply; consequent pressure would cause a decline in the population and the monotonous cycle would start anew: progressive development would momentarily arise only to be crushed by hunger and poverty. For the sober parson Malthus, this quelled the intoxicated hopes of the French Revolution. Yet Darwin found precisely in that situation a cycle, not of despair but of progressive alteration. A growing population would cause a struggle among members for available resources. Those organisms that by chance had traits that gave them an advantage would more likely survive the struggle, mate, and pass on their beneficial traits to offspring. The pressure, in Darwin's metaphor, would create a kind of filter, allowing only those better adapted to pass through.

> One may say there is a force like a hundred thousand wedges trying force into every kind of adapted structure into the gaps in the oeconomy of Nature, or rather forming gaps by thrusting out weaker ones. The final cause of all this wedging, must be to sort out proper

structure & adapt it to change—to do that, for form, which Malthus shows, is the final effect (by means of volition) of this populousness on the energy of Man.[110]

In this response to Malthus, Darwin displayed several elements, not quite analytically separated, of his inchoate device of natural selection. He recognized, for instance, that variable members of a population would mean that some individuals by chance would exhibit traits that more readily fitted them to their environment—an idea that would form one of the essential aspects of natural selection. In the last line of the quotation, he remarked another feature that Malthus's essay suggested to him: an enlarged population with scarce food supply would cause some individuals to decide to migrate to less densely populated areas. The analogy Darwin constructed was subtle and striking: he made human intentional action (i.e., decision to migrate)—which he regarded as a final cause—comparable to a final cause in nature. In nature, the result would be both the sifting out of the better endowed organisms and the dispersal of organisms into less densely packed territory. But both the sifting out and the dispersal would be the effect of a final cause, a cause that retained features of its primary analogue, namely, mental properties. The conception of final, mental-like causes in nature would persist through the construction of Darwin's theory.

The Origins of Moral Behavior

Almost immediately after the Malthus episode, Darwin began thinking seriously about human beings. Lamarck and others treated man no differently than other animals: humans, the Frenchman argued, derived from apelike ancestors. In order to bring man under the purview of his own transmutation theory, Darwin had to explain the salient feature of the human animal—not intelligence, but moral capacity. From the standpoint of the British empiricist tradition, of which Darwin certainly was a part, intelligence and reasoning were not unique to human beings. If intelligence amounted to the association of ideas, and ideas were copies of impressions, then no animal was barred from enjoying a modicum of rationality. But no animal, it was assumed, made moral choices. Lest his theory be brought back to the arena of special creation, Darwin had to show how moral behavior had its roots in man's animal predecessors. Only a few days after he read Malthus, Darwin started to construct a theory of moral evolution. On October 3, 1838, he jotted in his *Notebook N*:

> Dog obeying instinct of running hare is stopped by fleas, also by greater temptation as bitch . . . Now if dogs mind were so framed that he constantly compared his impressions, & wished he had done so & so for his interest, & found he disobeyed a wish that was part of his system, & constant, for a wish which was only short & might otherwise have been relieved he would be sorry or have troubled conscience. — Therefore I say grant reason to any animal with social & sexual instincts & yet with passion he *must* have conscience — this is capital view. Dogs conscience would not have been same with mans because original instinct different.[111]

In this entry, Darwin proposed that the moral impulse was a persistent and enduring instinct, which might come into conflict with other more abrupt and momentary urges. He quickly assumed that constant habits of a social character would provide the drive required for moral motivation, thus those of parental care, concern for the welfare of associates, and so on. Though his developing moral theory came in the wake of his initial formulation of the principle of natural selection, he did not attempt to account for moral instincts using the principle. One stumbling block to the application of natural selection was the problem of the recipient of advantage: moral action appeared to give benefit not to the actor but to the receiver. But natural selection, in Darwin's construction, operated to select traits that gave the possessor some advantage; moral traits thus seemed precluded from a natural selection explanation. Darwin reverted to the inheritance of habit for his account. It was only when he solved another problem in his developing theory, a problem about the social insects, did he see his way to a natural selection explanation of moral behavior.

Natural theologians had frequently looked to the behavior of insects as a sign of divine intervention. Darwin read with great interest the theologically cast *Introduction to Entomology* by William Kirby and William Spence. In his estimation they produced "the best discussion of instincts ever published."[112] These authors described in minute detail the structure of bee hives, the various casts of bee, from nurses to soldiers, and the large repertoire of wonderful instincts these creatures exhibited. The workers, for example, could sculpt perfect hexagonal cells out of the hive wax, and the soldiers would give up their lives to defend the hive. Such instinctive actions, in the view of Kirby and Spence, could only be the result of "faculties implanted in their constitution by the Creator."[113] Natural selection, again, seemed precluded, and this for several reasons: workers labored not for themselves but for the hive, and soldiers made the ultimate sacrifice for the common good. But beyond these considerations another seemingly insurmountable barrier loomed: workers and soldiers were neuters; they could not transmit advantageous traits to their offspring since they had none. Only in the throes of composing the manuscript of the *Origin* did Darwin hit upon the solution to his conundrum: natural selection didn't work on the individual, but the whole nest or hive. Those hives that by chance had, say, more aggressive individuals, would more likely survive a marauding bear than those hives lacking such protectors. The preserved hive could thus produce young queens carrying traits similar to those of their neuter sisters, the defenders; and those young queens might establish new hives and thus reproduce soldiers with augmented defensive instincts—and the cycle would continue. When he wrote the *Descent of Man*, Darwin would use the model of the social insects to explain the origin of moral instincts, instincts that worked, not for the benefit of the individual possessing them but for the individuals receiving the benefit.[114] This particular application of natural selection lay in the future. In the early period, as Darwin was thinking about the trajectory of nature, he understood that the moral dimension was crucial if human beings were to be given an account in terms of his theory.

The Teleological Trajectory of Nature

An even more salient employment of a final cause occurred when Darwin attempted to puzzle out the problem of sexual generation. On the initial page of his first transmutation notebook, *Notebook B*, Darwin wondered why asexual generation—the reproductive technique of plants, hydrozoa, sponges, simple worms—were not the dominant mode. Why did sexual generation arise? His initial speculation was that it had to do with adaptation to a changing world—"therefore final cause of life."[115] After more than a year of puzzling about this question, and after he had read Malthus, Darwin construed a deeper reason for sexual generation, one that displays clearly the pattern of his teleological thinking:

> My theory gives great final cause «I do not wish to say only cause, but one great final cause» . . . of sexes . . . for otherwise there would be as many species, as individuals, & . . . few only social . . . hence not social instincts, which as I hope to show is «probably» the foundation of all that is most beautiful in the moral sentiments of the animated beings.[116]

The argument is classically teleological: sexual generation exists in order that stable species (instead of mere individuals) should exist; and stable species exist in order that social animals should exist; and social animals should exist in order that social instincts should exist; and social instincts should exist in order that moral animals might come to exist. In this pattern of explanation, the antecedent condition occurs because of its consequent, its goal or end. The next line in the quotation makes explicit the final cause of the whole series: "If man is one great object, for which the world was brought into present state . . . & if my theory be true then the formation of sexes rigidly necessary."[117] In Darwin's view, the world and all of its organisms were made for man. Human beings, the goal of nature, made the antecedent condition, sexual generation, "rigidly necessary." How deeply this penetrated into Darwin's thinking about nature can be gauged by the archeological structure of his theory.

Consider the final passage of Darwin's *Origin of Species*. He first mentions the laws that his theory put on display: variability, inheritance, struggle for existence, and natural selection. Then he concludes that from these laws, "the most exalted object which we are capable of conceiving, namely, the production of the higher animals directly follows." He finishes with that lyrical exultation:

> There is grandeur in this view of life, with its several powers, having been originally breathed into a few forms or into one; and that, whilst this planet has gone cycling on according to the fixed law of gravity, from so simple a beginning endless forms most beautiful and most wonderful have been, and are being evolved.[118]

But how are we to understand the passage? What produces the grandeur? I think the structure of the passage and its teleological import can be understood by digging back through the antecedent passages that are variations on the common structural scaffold for the *Origin*.

1. "The most exalted object which we are capable of conceiving, namely, the production of the higher animals directly follows" —*Origin of Species* (1859).
2. "The most exalted end which we are capable of conceiving, namely the creation of the higher animals, has directly proceeded" —"Essay of 1844."
3. "The highest good, which we can conceive, the creation of the higher animals has directly come" —"Essay of 1842."
4. "Man is one great object, for which the world was brought into present state" —*Notebook E* (1838).

That final peroration of the *Origin* is a teleological invocation: the purpose of the laws of nature—their "object"—is to produce the higher animals, including the highest—human beings. Darwin feigned no mere rhetoric when he exclaimed to Asa Gray: "I am inclined to look at everything as resulting from designed laws."

I don't claim that Darwin was fully reflective about the deeper structural character of his theory. But if his theory resides in the words he wrote, then the teleological structure is patent. But this conclusion can be approached from another angle, namely, if we examine more minutely Darwin's construction of the principle of natural selection itself. I think it will become clear that nature, in the view of the theory, is guided by mind.

Natural Selection: Mind in Nature

Darwin met several obstacles when trying to formulate his incipient principle of natural selection. I've indicated one: the problem of the neuter insects. But there were other, even more fundamental difficulties: How can natural selection pick out minute traits and finely shape the structure of organisms? And how can it prevent those traits from being swamped out by disadvantageous or neutral traits? Huxley recognized the former problem, and advised Darwin to drop the notion of very small, gradual increments as the object of selection. Huxley, the mechanist, knew a machinelike principle could not act on exceedingly slight traits so as to produce virtually imperceptible improvements at each iteration. The latter problem is captured by that line from Stephen Stills: "If you can't be with the one you love, love the one you're with." But if you love the one you're with, that mate would not likely have the chance, advantageous trait and the resultant progeny would thus likely have it only in a reduced form. That is, when a favorable trait occurs infrequently and by chance, the individuals in your neighborhood with whom you might mate would probably lack the advantageous trait or have an unfavorable version of it, thus reducing the favorable trait in the next generation. Eventually the favorable variation would be swamped out—a problem Darwin keenly appreciated.

Darwin dealt with these difficulties at two related levels: at a theoretical level and at an empirical level. The theoretical level constrained and shaped what he took to occur at the empirical. First, Darwin needed some account of the selection of very small, almost imperceptible differences; and second, an account that would explain the segregation of favored organisms. These problems could be resolved if the natural process were comparable not to a machine but to a very powerful mind that could see into the very fabric of organisms, select minute and obscure traits, and preserve their carriers for mating—and then segregate those carriers from the rest of the group. These requirements were met by the model he introduced initially in his "Essay of 1842" to describe the actions of natural selection. That model was elaborated in his "Essay of 1844"—a model of a very powerful mind:

> Let us now suppose a Being with penetration sufficient to perceive differences in the outer and innermost organization quite imperceptible to man, and with forethought extending over future centuries to watch with unerring care and select for any object the offspring of an organism produced under the foregoing circumstances; I can see no conceivable reason why he could not form a new race (or several were he to separate the stock of the original organism and work on several islands) adapted to new ends. As we assume his discrimination, and his forethought, and his steadiness of object, to be incomparably greater than those qualities in man, so we may suppose the beauty and complications of the adaptations of the new races and their difference from the original stock to be greater than in the domestic races produced by man's agency.[119]

The essays constituted Darwin's attempt to sketch for himself the theory of natural selection that had slowly coalesced in his mind. There are several features of his proposal that need to be emphasized. First, this conception was not meant for other eyes, only for fixing the outlines of the theory in his own mind—that was certainly the case for the "Essay of 1842"; thus we should not suspect he was simply trying to assuage the concerns of a general public. Second, and most importantly, this is a model, not of a machine but of a very powerful intelligence. No machine could see into the very fabric of creatures, could detect very small, virtually imperceptible, variations for selection. No machine could segregate favored individuals and bring them together for mating. Third, the model could see into the future; it had foresight, as the above passage indicates. Fourth, it expressed concern for creatures and was thus altruistic, unlike "blind foolish man," who selected creatures not for their welfare but for his.[120] Fifth, though the powerful intelligence was not God, Darwin here suggested that it was acting as a surrogate for the Creator, a secondary cause. Finally, this model was not something Darwin constructed and then quickly abandoned. He retained the model in the *Big Species Book*—the manuscript for the *Origin of Species*—and in the *Origin* itself.[121]

When Darwin explained the actions of natural selection in the *Origin of Species*, this powerful intelligence again made its appearance:

> Man can act only on external and visible characters: nature cares nothing for appearances, except in so far as they may be useful to any being. She can act on every internal organ, on every shade of constitutional difference, on the whole machinery of life. Man selects only for his own good; Nature only for that of the being which she tends. . . . Can we wonder, then, that nature's productions should be far "truer" in character than man's productions; that they should be infinitely better adapted to the most complex conditions of life, and should plainly bear the stamp of far higher workmanship? It may be said that natural selection is daily and hourly scrutinizing, throughout the world, every variation, even the slightest; rejecting that which is bad, preserving and adding up all that is good; silently and insensibly working whenever and wherever opportunity offers, at the improvement of each organic being.[122]

Though the phrase "mechanism of natural selection" comes trippingly to our tongues—and Ruse uses the phrase reflexively when referring to Darwin's conception—it never rolled off Darwin's tongue, and for good reason. What Darwin described as the actions of selection could not be performed by any mechanism of his acquaintance, though the clan of Darwinian scholars has simply assumed that the English master introduced blind mechanism into nature. Consider again the traits of this extremely intelligent process: it operates on very small variations throughout the entire organism—beyond the capacity of any machine. The consequence was an "infinitely better adapted creature"; in *Big Species Book*, Darwin said the selections would be "perfect adaptations to the conditions of existence."[123] And in the *Origin*, Darwin made explicit the moral character of the natural process of selection.

In his description of natural selection, we should note that the process works for the good of the organism, unlike actions of the human breeder; that is, natural selection is an altruistic process, while human selection is selfish. Darwin was emphatic about this when he asserted, in the above passage, that natural selection works for "the improvement of each organic being." That's not a mere slip of the pen. Darwin used a phrase like that at least six times in the *Origin*.[124] In the penultimate paragraph of the book, he put it this way: "And as natural selection works solely by and for the good of each being, all corporeal and mental endowments will tend to progress towards perfection."[125] How different is Darwin's conception from our view. We do not assume that natural selection works "by and for the good of each being." In our view, natural selection destroys most beings, obliterates them. If Darwin were confronted with the logic of his own device of natural selection, he might well have retracted those several assertions. But he was so convinced of the moral and intelligent character of nature and its operations he failed to notice what seems to us the very logic of natural selection.

Another measure of the depth of Darwin's assumption that natural selection was an intelligent and moral process is the more general character of Darwin's language in the *Origin*. The term "mechanism" in any of its forms—mechanistic, mechanical, machinery—occurs only five times in the *Origin*, and none of those characterize the actions of natural selection itself. The term "purpose" and its more obscure synonym "object"—as in "my object here is to convince you"—occur some sixty-seven times.

Darwin's conception of natural selection had consequences for his understanding of the evolution of organisms. First, he conceived the temporal pace of evolution to be exceedingly slow and insensibly gradual. Though in the *Origin* he spoke of natural selection as "daily and hourly scrutinizing every variation," he estimated that selection of useful variations might occur only a few times "in the course of thousands of generations."[126] Only an intelligent force might act with such deliberation and patience. Huxley, who did believe nature acted in stuttering, mechanical fashion, thought Darwin made a mistake in claiming that selection produced only very slow, gradual changes. "Mr. Darwin's position might, we think, have been even stronger than it is," Huxley complained, "if he had not embarrassed himself with the aphorism, 'Natura non facit saltum' which turns up so often in his pages."[127]

A second consequence of the way Darwin construed natural selection regarded the environment of its operation: it worked most efficiently in large open areas, selecting scattered individuals for mating, much as the successful nurseryman kept large numbers of plants or animals so that a few favored individuals might be more easily gathered for reproduction. Darwin simply supposed that a virtually intelligent natural selection could mitigate the swamping problem by somehow segregating the best individuals and preventing backcrosses to those that lacked the selected traits. When Fleeming Jenkin, in a sharply critical review of the *Origin*, insisted on the problem of swamping, Darwin recognized that his device really couldn't escape the difficulty.[128] In the fifth edition of the *Origin*, he responded to Jenkin's objection by invoking a Lamarckian scenario, a kind of deus ex machina. He proposed that in a large, extended environment, all of the plants or animals in a given subregion might vary together, so that selection could occur without the danger of swamping.[129] Though natural selection had lost its intelligent power in this emendation, Darwin yet had to arrange for a semimiracle to respond to Jenkin's objection. Today, most evolutionary biologists doubt the kind of *sympatric* speciation Darwin assumed; they hold that selection more easily operates on small breeding groups separated from other groups by physical barriers, what is now called *allopatric* speciation.

A third consequence of Darwin's conception of natural selection led to a puzzle in the *Origin of Species*: Why are there two chapters on natural selection? The third chapter is entitled "Struggle for Existence" and the fourth "Natural Selection," as if these were different things. In the 1842 and 1844 essays, Darwin introduced the language of the intelligent and moral selector first, then gave only a few sentences to the idea of struggle for existence; all of this fell under the title "natural selection." In the original manuscript for the *Origin*, published in modern times as *Big Species Book*, Darwin began chapter 5 under the title "On Natural Selection"; but then retitled the chapter "The Struggle for Existence Bearing on Natural Selection" and followed it with chapter 6, "On Natural Selection," which latter included the language of the intelligent and moral selector.[130] In the essays, Darwin seemed very concerned with the swamping problem, which an intelligent selector might deal with. As struggle came to be seen as the natural instantiation of the operations, via natural law, of a powerful mind, Darwin placed the chapter on struggle just after the chapter on variation in nature (chapter 2 of the *Origin*). If the language regarding the intelligent and moral selector were only rhetorical and dispensable in Darwin's theory, then he might well have eliminated the language altogether. Indeed, Wallace had suggested to Darwin in 1866 that the term "natural selection" was too anthropomorphic and that he should replace it with Herbert Spencer's phrase "survival of the fittest."[131] In the fifth edition of the *Origin* (1869), he did retitle his chapter on natural selection, calling it "Natural Selection or the Survival of the Fittest." He thus did not eliminate the term or his language of the intelligent selector—so intimate a part of his conception did the model of the powerful mind play. Wallace's suggestion now seems strangely redundant. After all, "struggle for existence" is hardly different from "survival of the fittest"; neither carries the suggestion of an intelligent selector.

If natural selection were a wise and morally inclined governor of the evolutionary process, then it ought to be, as Darwin assumed it was, an instrument of progressive development: creatures in subsequent generations, he supposed, would be "higher" than those of earlier generations. Progressive traits of organisms accumulated over time. So, in Darwin's estimation, if a more advanced animal were placed in an earlier environment, it would be favored over those indigenous competitors.[132] The progressive character of evolution would be necessary if the goal were to produce the higher animals, especially the moral animals. Darwin had conceived natural selection as precisely the kind of agent that would continuously produce cumulative improvement: "And as natural selection works solely by and for the good of each being, all corporeal and mental endowments will tend to progress towards perfection."[133]

A fourth consequence of Darwin's model of natural selection—which includes his version of community selection—results from his assumption that selection has a moral character: natural selection operates on the group without requirement that members be related. Ruse has long held that Darwin was not a "group selectionist" but only an individual selectionist.[134] For Ruse, individual selection just seemed more in keeping with a laissez-faire British impulse. To maintain this view, he has to suppose that Darwin's conception of community selection is a kind of individual selection rather than group selection. Would Darwin have recognized the distinction? The problem of group selection—and thus the explicit boundaries of the concept—arose only in the second part of the twentieth century. While Darwin did discriminate groups in which most of the members were related from groups in which most were not, he seems oblivious to the kind of difficulty brought to light by George Williams and Richard Dawkins a hundred years later. And, of course, he gave no hint of being presciently privy to Hamilton's formulation of the idea of inclusive fitness.[135] The measure of Darwin's move to group selection can be taken from a passage that underwent continual alteration in subsequent editions of the *Origin*. In the first edition the passage read: "In social animals it [natural selection] will adapt the structure of each individual for the benefit of the community, if each in consequence profits by the selected change."[136] The "each" in the second part of the proposition refers to individuals making up the community. But in the fifth edition of 1869, the last phrase in the sentence was altered to read: "if this [i.e., the community] in consequence profits by the selected change." And in the last edition, Darwin removed any ambiguity: "In social animals it [natural selection] will adapt the structure of each individual for the benefit of the community; if the community profits by the selected change."[137] The fifth and sixth editions of the *Origin* (1869 and 1872) were published just before and just after the *Descent of Man* (1871), wherein Darwin deployed the concept of community and group selection to explain both the rise of human intelligence and man's moral capacity (discussed below). Darwin becomes an individual selectionist only if one turns him into a twenty-first-century biologist. It seems to have escaped the notice of some scholars that Darwin lived in the nineteenth century.

A final consequence of Darwin's conception of natural selection unites his beginning with his end, or rather the end of the *Origin*. In the *Notebooks*, Darwin had conceived man as the "one great object, for which the world was brought into present state"—man that highest animal, man the moral animal. At the conclusion of the *Origin*, in that passage which he began honing in the 1842 and 1844 essays, Darwin declared that from the laws he had discriminated, especially natural selection, there came "the most exalted object we are capable of conceiving, namely the production of the higher animals." And of course, as he had charted from the beginning, that highest animal was the one with moral sentiments, man. There was indeed, "grandeur in this view of life."

Darwin's conception of natural selection as an intelligent and moral force, with the Creator receding into the background, had consequences. The garden of nature was not replaced by Locke's spinning-jenny; it rather flourished as cultivated by that superior Romantic gardener, Charles Darwin. Later, it is true, Darwin's enthusiasm for mechanism grew, but his more organic, more moral conception of natural selection left a lasting imprint on the *Origin of Species*.

Natural Selection from Individual to Group Selection

In order to make Darwin's commitment to group selection perfectly evident, let me briefly rehearse the development of his thought on the issue. In the first edition of the *Origin*, he recognized two rather different kinds of natural selection: individual selection, which operated on the traits of the individual; and family or community selection, which, in the case of the social insects, operated on the whole hive of bees or nest of ants. Darwin was fully aware that in this latter case, the members of the community were related.

In 1864, before the Anthropological Society of London, Alfred Russel Wallace delivered a lecture on the origin of the human races as explained by natural selection. In his paper, published in the same year, he developed two arguments that made a considerable impact on Darwin.[138] He proposed that at a certain point in the evolution of primitive anthropoids, natural selection ceased to act on the bodies of these creatures but continued to act on their minds. This would explain the similarity in bodily form of the great apes and human beings, and, as well, their great differences in intelligence and moral capacity. Darwin confessed to Wallace that this idea was novel to him, namely, "that during late ages the mind will have been modified more than the body."[139] Wallace's second argument specified the mode of selection on intelligence and moral behavior:

> Capacity for acting in concert, for protection and for the acquisition of food and shelter; sympathy, which leads all in turn to assist each other; the sense of right, which checks depredations upon our fellows [etc.] . . . are all qualities that from their earliest appearance must have been for the benefit of each community, and would, therefore, have become the subjects of "natural selection." . . . Tribes in which such mental and moral qualities were dominant, would therefore have an advantage in the struggle for existence over other tribes in which they were less developed, would live and maintain their numbers, while the others would decrease and finally succumb.[140]

Darwin employed almost these very phrases in giving his own account, in the *Descent of Man*, of human intelligence and morality. Though I will follow out Darwin's theory of the evolution of these traits in the next section, let me simply quote from the *Descent of Man* to show the very close connection between that account and Wallace's 1864 paper, which Darwin so admired. Here's Darwin in the *Descent*:

> There can be no doubt that a tribe including many members who, from possessing in a high degree the spirit of patriotism, fidelity, obedience, courage, and sympathy, were always ready to give aid to each

other and to sacrifice themselves for the common good, would be victorious over most other tribes; and this would be natural selection.[141]

What is striking about Wallace's and Darwin's respective accounts of morality is that both of them apply natural selection to the whole community and that they do so using virtually the same language and without regard to any familial relationships of members. Indeed, Wallace, many scholars have supposed, originally spoke in terms of selection operating on groups of individuals rather than directly on the individuals themselves.[142] And so, I believe, did Darwin, at least when describing the evolution of intelligence and morality. This development of his ideas explains, I believe, those altered sentences in the later editions of the *Origin*, which I quoted above, where it becomes obvious that he has allowed for group selection of social communities. In the following section, I will provide further evidence of Darwin's commitment to group selection and will show why, analytically, he could not have been anticipating anything like Hamilton's theory of inclusive fitness.

MAN, THE MORAL ANIMAL

Darwin did not leave man morally naked to the world. The *Origin of Species* portrayed nature as having a goal, the production of the higher animals, of which the highest were human beings with their moral sentiments. From the earliest period of his theorizing, Darwin recognized that he would have to give an account of human beings, particularly a developmental explanation of their capacity for moral reasoning and behavior. There is evidence that he had intended to treat human beings in the *Origin*, but he seems to have thought readers would be transfixed by that subject, allowing deep prejudice and ancient orthodoxy to smother evidence and reason. Only one passing sentence explicitly considered the human animal, and that in the last pages of the book: "Light will be thrown on man and his history."[143] Darwin's caution was misplaced. The first reviewers of the volume understood immediately what the theory implied about human beings. In an early review, for example, Adam Sedgwick, who took the young Darwin on a geologizing trip before his embarkation on the *Beagle*, thought Darwin's theory in the *Origin* implied humans were "nothing better than the natural progeny of a beast."[144] Sedgwick objected that since Darwin did not discuss human faculties of moral capacity, the theory regarded man no better than a wild animal, "which has to live, to beget its likeness, and then die for ever."[145] In response to his critics, like Sedgwick, Darwin resolved to treat human beings in his *Variation of Animals and Plants under Domestication*—appropriate enough, since, as he slyly told Wallace, man seemed "an eminently domesticated animal."[146] His own work on *Variation*, however, went to two volumes, and he didn't think he could do justice to human beings in just a chapter. He was going to treat the evolution of the human races from the point of view of sexual selection, but then he got into a row with Wallace about that kind of selection in birds. The topic ballooned into a long treatise on the subject. Then the unthinkable happened. Wallace defected.

Wallace, during the latter part of the 1860s, became interested in the phenomenon of spiritualism. Naive soul that he was, he became convinced that a spirit world existed and that gifted individuals could communicate with denizens of that dimension—even beyond that, could photograph them. Wallace had his picture taken during a séance conducted by the celebrated medium Mrs. Agnes Nichol Guppy; on development of the photo, an ectoplasmic spirit materialized right beside Wallace (figure 20). Sherlock Holmes's creator, Arthur Conan Doyle, also saw spirits in photographs shown to him, though in his case they were fairies. Darwin was a bit disappointed with his own experience attending a séance. The spirits stayed out of sight and mum.

Wallace became a believer, especially when the existence of a spirit world explained for him certain features of human evolution for which natural selection seemed inadequate. He wrote Lyell in April 1869 to make his objection explicit: man's big brain, his ethical capacity, his naked skin, and his musical ability could not be accounted for by natural selection because these traits had no use, no survival value. "Unless Darwin," he asserted, "can shew me how this rudimentary or latent musical faculty in the lowest races can have been developed by survival of the fittest — can have been of use to the individual or the race, so as to cause those who possessed it in a fractionally greater degree than others to win in the *struggle for life* I must believe that some other power caused that development."[147] Wallace presumed that higher powers selected humans who had desirable traits much in the way we chose animals for their special properties. Darwin was aghast.

Wallace's defection on man, as well as the growing pressure to bring humans under the aegis of his theory, moved Darwin to turn his manuscript on sexual selection into a far-ranging discussion of human evolution, especially of those distinctively human qualities that Sedgwick and Wallace thought only higher powers could explain. In the resulting two-volume *Descent of Man and Selection in Relation to Sex* (1871), Darwin confined the problems of sexual selection to the second part, which defined the original scope of his intended book. In that second part, Darwin argued that the distinctive traits distinguishing males from females in the animal kingdom could largely be explained by sexual selection (i.e., through male combat or female choice). In the case of the human animal, sexual selection might account for racial differences in skin color, hair, and body types, as well as differences in male and female secondary sexual characteristics, body structure, and mental features (e.g., intelligence, vocal ability, and tender virtues). The first volume of the *Descent* was more hastily constructed. Darwin briefly discussed the animal antecedents for several human psychological and mental traits, but saved his extended considerations for just those features of human beings that Wallace assigned to the provenance of higher spiritual powers, especially an enlarged moral sense. In the following sections I will describe Darwin's accomplishment in this regard. Darwin provided, I believe, a compelling analysis of human moral judgment. Some contemporary scholars have offered a quasi-Darwinian account of morality, but not as an expression of an authentic morality. They rifle from Darwin only the flimsy dress of morality, a faux morality; moral judgment, they argue, cannot be objective. Darwin certainly believed it could be objective. The taste for moral behavior, these neo-Darwinians suppose, must be like a taste for cured olives, simply an individual, subjective preference. I will dispute that conclusion.

In the first part of the *Descent of Man*, Darwin suggested what might be the animal roots of the various qualities conspicuous in humans: imagination, memory, dreaming, aesthetic attractions, play, language, and other traits. Few naturalists of his time and place would doubt that the rudiments of these traits could be found in animals. Not all, to be sure, were perfectly sanguine about what seemed to be Darwin's credulity concerning animal abilities. A pointed example came from a story he drew from Brehm's *Thierleben*. It was a tale about a baboon that had adopted a stray kitten, but was surprised when it got its nose scratched by tiny claws. The baboon, in Darwin's retelling, was astonished, and being intelligent and of a practical mind, seized the kitten's feet and bit off its claws. St. George Jackson Mivart, in an anonymous review of the first edition of the *Descent*, dismissed the story with a sneer, saying it was perfectly impossible for the baboon to have clipped off the kitten's claws.[148] In the second edition of the *Descent*, Darwin, ever the empiricist, assured his readers in a note: "I tried and found that I could readily seize with my own teeth the sharp little claws of a kitten nearly five weeks old."[149] Despite some dubious tales of animal intelligence, no British empiricist (or dedicated hunter) doubted that animals had a portion of reason in common with human beings. The case was otherwise concerning morality. That was the Rubicon that no animal could cross.

So Darwin had the task of discovering the rudiments of morality in animals and then developing a plausible theory of its evolution in human beings. His immediate critics were severe in their reaction, and many simply misunderstood the import of his analysis. Wallace had already assigned human acquisition of moral capacity to higher powers. The reviewer of the *Times* thought that Darwin was "quite out of his element" in treating mind and morals.[150] The *Edinburgh Review* sounded an alarm still heard loudly today in religiously conservative circles:

If our humanity be merely the natural product of the modified faculties of brutes, most earnest-minded men will be compelled to give up those motives by which they have attempted to live noble and virtuous lives, as founded on a mistake. . . . If these views be true, a revolution in thought is imminent, which will shake society to its very foundations, by destroying the sanctity of the conscience and the religious sense; for sooner or later they must find expression in men's lives.[151]

Despite the reviewer's warning alarm, the revolution has occurred, at least within the provinces of the academy. Most biologists and humanities scholars have accepted the basic terms of Darwin's theory of human evolution, though keen disputes do remain over an evolutionary construction of moral behavior. I will consider these various contemporary interpretations momentarily. Let me first turn to Darwin's theory of the evolution of intellect.

Darwin's Theory of Intellectual Evolution

In the *Descent of Man*, Darwin answered the several objections of Wallace concerning the suite of apparently useless but distinctive traits exhibited by human beings. Man's naked skin, for example, Darwin attributed to sexual selection.[152] He explained human racial and dimorphic characters also through sexual selection. Wallace thought human beings displayed intelligence far beyond what was needed for simple survival—thus inexplicable by natural selection. Darwin, in response, provided a complex account of human intellectual ability that bespeaks the layered composition of his book.

In the second volume of the *Descent*, Darwin explained the advance in human intelligence through the actions of natural and sexual selection (e.g., the cunning required for male combat).[153] Yet, he knew that would not quite satisfy Wallace's objection, so he came at the topic again in the first volume of the book. There he claimed two other sources of accelerated brain power among proto-humans. First would be the effect of language usage. Darwin argued that as language complexity increased due to human interchange, brain would be increasingly exercised in reciprocal fashion; thus mind, the manifestation of brain, would grow in power and sophistication: "for the continued use of language will have reacted on the brain, and produced an inherited effect; and this again will have reacted on the improvement of languages."[154] So exercise of the muscles of the brain, through ever more complex modes of language, might shove proto-humans into humanity. This quasi-Lamarckian explanation did not fall prey to Wallace's trap concerning useless traits. Darwin, though, was rarely satisfied with one explanation of a phenomenon. If the reader proved skeptical concerning the impact of language, Darwin had another account perched at the ready: community selection.

He observed that within a tribe of proto-humans, if an individual by chance had a superior intellect and constructed some invention, others taking up the invention through imitation would exercise their brains in a way comparable to language learning. Their children would inherit this capacity, and the tribe would flourish; and so by reason of the inheritance of structures modified by exercise, tribal groups would increase overall mental ability. But Darwin then added a natural selection feature to this proposal:

> Even if they [i.e., those accidental geniuses] left no children, the tribe would still include their blood-relations; and it has been ascertained by agriculturists that by preserving and breeding from the family of an animal, which when slaughtered was found to be valuable, the desired character has been obtained.[155]

Darwin was thus able to overcome one of Wallace's principal objections. His account, though it initially employed the inheritance of acquired characteristics, quickly became transformed into one of community selection: those small tribal groups that had by chance a genius whose inventions spread through the tribe—that tribe would have an advantage over other tribes, and so would be naturally selected. The tribe so selected would likely flourish and would advance intellectually through continued competition over many generations.

Is Darwin's use of community selection of intellect an example of individual selection or of group selection? As noted above, Ruse has assimilated Darwin's theory of community selection to contemporary notions of a kind of kin selection; he does so in an effort to maintain Darwin as an individual selectionist and to show that he had not succumbed to Romantic and airy notions of group selection. But as I suggested above, this is the kind of move—making Darwin's theory virtually the same as our own—that does a historical injustice to the great nineteenth-century thinker. And we need to pay attention to the logic of the situation.

In the above quotation, Darwin certainly does mention that the tribe which receives the benefit from the inventive individual and which consequently succeeds in competition with other tribes might have his relatives also receive the benefit. It almost looks like a matter of kin selection, but we need to pay attention to the logic of Darwin's conception. First, the benefit of the genius's invention, under Darwin's scenario, extends throughout the tribe, not directly benefiting each individual, but only the group as a whole; because the group has the advantage in competition with other groups, it will have a better chance of survival, which would indirectly benefit both the genius's relatives and other nonrelatives. This is not a Hamiltonian model of kin selection, according to which it is only the relatives of a self-sacrificing individual who benefit and *not* other members of the same community. Darwin made it clear that the focus of natural selection in these instances was not the individual members of the community seriatim, but the community at large:

> These [mental] faculties have been chiefly, or even exclusively, gained for the benefit of the community; the individuals composing the community being at the same time indirectly benefited.[156]

It was this conception, almost exactly in these words, that was expressed in the emendation of the relevant passage in the later editions of the *Origin* (mentioned above). If the advantage is principally to the whole community, then that must be the object of selection's grasp, with individuals, both related and unrelated, only indirectly receiving the benefit.

Darwin was just as likely to provide a group selectionist account of the rise of intellect without any mention that the group needs be composed of related members. So, for example, in speaking of the evolution of humans on the remote and warm islands of the South Pacific, protected from most predators (e.g., on Borneo), he maintained it would simply be the competition among tribes that would drive higher intellect:

> In an area as large as one of these islands, the competition between tribe and tribe would have been sufficient, under favourable conditions, to have raised man, through the survival of the fittest, combined with the inherited effects of habit, to his present high position in the organic scale.[157]

This passage seems to imply that natural selection would preserve one tribe rather than another because of its advantageous trait (i.e., imitating the genius's discovery), and that kind of preservation would allow inherited habit to elevate that tribe's descendants. Perhaps, as well, the genius's relatives would also be preserved and be able to pass their improved mentality to succeeding generations. There are some problems with this scenario, of which Darwin seems not to have been aware. The crucial point is, however, that he introduced a group selection solution to the rise of human intellect. His commitment to a group selection hypothesis becomes even more obvious in his theory of the evolution of morality.

Darwin's Theory of Moral Conscience and Behavior

As I have already indicated, Darwin recognized he would have to give an account of human moral development lest his theory be burked by the intervening hand of the Creator. During the period prior to the composition of the *Origin*, Darwin presumed that moral judgment was based on an innate instinct for right action. He held that such impulses had roots in the social and parental instincts of animal predecessors. In his early notebooks, Darwin proposed a Lamarckian kind of account for the origin of such instincts. Only in the latter stages of his theorizing did he begin to apply natural selection to instinctive behavior, and only after he discovered community selection did he have the right instrument to formulate a natural selection theory of moral instinct.

In the first volume of the *Descent*, Darwin declared that the most marked difference between man and the lower animals was the moral sense. While he recognized that many writers of consummate skill had written on human conscience, he thought he might venture something from the side of natural history. Chief among those other writers was Immanuel Kant, whom Darwin briefly and surprisingly quoted:

> Duty! Wondrous thought, that workest neither by fond insinuation, flattery, nor by any threat, but merely by holding up thy naked law in the soul, and so extorting for thyself always reverence, if not always obedience; before whom all appetites are dumb, however secretly they rebel; whence thy original?[158]

Darwin, of course, thought he knew duty's "original." He had been urged to read Kant by Frances Power Cobbe (1822–1904), a woman of considerable intellectual ability whom the family met while visiting London in 1868. Darwin initially resisted Cobb's urging that he read the English translation of Kant's *Metaphysical Foundations of Morals*, which she sent him.[159] But he did relent. He got far enough into the book to appreciate how different his own approach would be.[160] Likely, though, he felt compelled to flesh out a bit more the conditions under which the social instincts might be turned into judgments of conscience and human moral behavior.

Darwin specified the conditions under which conscience in proto-human beings would develop.[161] First, there had to be a repertoire of social instincts. Darwin described these instincts as directed to the welfare of others, that is, "to aid their fellows in certain general ways."[162] I will return to this requirement in a moment, since it is the most fundamental. The second condition he specified was sufficient intellect to recall when a social impulse had not been requited, but perhaps overcome by a momentarily stronger impulse of a self-regarding kind. This condition rehearsed the fundamental idea Darwin laid down in his early notebooks, namely, that the social instincts were persistent and of longer duration than fleeting, self-regarding instincts. Darwin suggested that a mother bird with sufficient intellect (of the human variety), if she had abandoned her young to fly south in the winter, when she reflected on that more persistent urge to care for her young, she would have a troubled conscience—hence, the pangs of guilt upon violation of a deeply held moral rule. The third requirement would be language, which would enable members of the community to express their needs. This would allow the social instincts to impel action "for the good of the community."[163] Finally, in his Lamarckian-like way, Darwin believed that the social instincts would be strengthened by habitual exercise.

The first requirement, the establishment of innate drives to work for the community good, was fundamental for Darwin's conception of morals, and he spent some time in elaborating how these instincts would be acquired. As a first stage, still when the human stock had not appeared, the social instincts would likely be deeply rooted as family values; the parental instincts would be an anchor. But as proto-humans gradually achieved their humanity, the focus of natural selection would transcend the family and become directed to the tribal community. The fundamental process became community selection:

> There can be no doubt that a tribe including many members who, from possessing in a high degree the spirit of patriotism, fidelity, obedience, courage, and sympathy, were always ready to give aid to each other and to sacrifice themselves for the common good, would be victorious over most other tribes; and this would be natural selection. At all times throughout the world tribes have supplanted other tribes; and as morality is one element in their success, the standard of morality and the number of well-endowed men will thus everywhere tend to rise and increase.[164]

This passage is obviously one of group selection, not of kin selection or family selection: there is no suggestion that members of the community should be constituted only by relatives. As Darwin made clear: "These [mental] faculties have been chiefly, or even exclusively, gained for the benefit of the community."[165] This passage appears to be virtually the same as the emended passage in the last edition (1872) of the *Origin of Species*: "In social animals it [natural selection] will adapt the structure of each individual for the benefit of the community; if the community profits by the selected change."[166]

Note that in the long passage quoted above, Darwin claims that those societies having more individuals displaying "patriotism, fidelity, obedience, courage, and sympathy" would have the advantage over those societies lacking such individuals; but having more such altruistic individuals can only be a property of the group, since only groups can have *more members*. This logical feature of Darwin's scenario needs to be emphasized, since it quite clearly distinguishes his proposal from any comparable proposal based on similarity to kin selection. First, Darwin here describes a feature characteristic of the group at large—that is, more members that are cooperative and self-sacrificing; and, again, *more members* can only be a trait of groups in competition with other groups, not a trait of an individual. Individuals would indirectly benefit, as Darwin noted; but the unit of natural selection's action could only be the group. Second, in kin selection, an altruistic trait, which is disadvantageous to the individual, will be selected if the trait sufficiently benefits the individual's relatives; but in group selection, a trait characteristic of the whole group will be selected if it benefits the group as a whole, not particular individuals within the group. Darwin perhaps thought that family members in a society who indirectly benefited from a relative's invention or altruistic behavior would be the hereditary reservoir to carry the innate instincts into the next generation. Whether such a circumscribed pool of heredity would be enough to continue the evolution of morality or intellect seems to me uncertain. Darwin was not privy to all the problems attendant on group selection that were recognized only in the mid-twentieth century. He nonetheless attempted a group selection solution to the problem of human altruism, and that is the point to be made.

Darwin supposed that in order for community selection to get a firm foothold there might be initiating conditions that promoted actions that appeared to be altruistic and selflessly cooperative. These conditions would be based in selfishness but might yet promote apparently altruistic actions: reciprocal altruism and praise and blame would be the means. Darwin thought both of these functioned as a "low motive."[167] The first assumed that if one helped another, aid might be returned in the future; this was reciprocal altruism *avant la lettre*. The second would occur when members of a tribe praised others for acts of cooperation or courage; that would have the effect of increasing the incidence of such acts. At times, Darwin suggested that reciprocal altruism and praise and blame might be initiating causes for cooperative or self-sacrificing behavior among tribes of proto-men. At other times, he appears to have regarded such causes as auxiliary to the main cause, since he recognized that praise and blame could have effect only if the unselfish social instincts had already been established.[168] Nonetheless, both of these causes would, in Darwin's view, be low motives, since they would be performed for ultimately selfish reasons: the hope of future reward or the gratification that praise brings. But Darwin was absolutely clear that these were not of the essence of his theory of morality, since his theory based morality on authentic altruism—the social instincts—without taint of selfishness; and these were established through community selection. Reciprocal altruism and praise and blame would be based on a desire for one's own good; they would be secondary causes at best:

> As all wish for happiness, the "greatest happiness principle" will have become a more important secondary guide and object; the social instincts, including sympathy, always serving as the primary impulse and guide. Thus the reproach of laying the foundation of the most noble part of our nature in the base principle of selfishness is removed.[169]

Though many critics and even sympathetic scholars caricature a Darwinian approach to ethics as based on selfishness, the master himself certainly did not make a similar judgment.

In the first part of this essay, I have attempted to demonstrate the influence on Darwin's thought of both the British tradition of natural theology and the leading principles of German Romanticism. In his conception of the evolution of morality, one can detect these same influences at work, if a bit more indirectly. Paley had grounded moral judgment in utility, that vague concept stemming from Hume and Bentham, which had at its core the assumption that pleasure and pain were the drivers of all action. Darwin yet recognized, perhaps abetted by his cursory acquaintance with Kant, that authentic morality could not be selfishly motivated.[170] He was not alone in moving away from the tradition of Jeremy Bentham and James Mill. John Stuart Mill, for instance, struggled to wring disinterested moral judgment from individual judgments of utility.[171] Even if the transition in Mill's theory lacked the necessary logical links, this reformed utilitarian saw what was required; and his study of Coleridge helped him to be so aware. In his marvelous essay on Coleridge, Mill got the German perspective on morality—which the poet gently lifted from Kant and Schelling—as something innate in human beings. Darwin read Mill's anonymously published essay in 1840 and, of course, found the notion of an innate moral impulse quite congenial to his developing theory.[172] Darwin's own theory was thus not on the side of the British empiricist tradition but on that of Coleridge and his German guides. Darwin made this clear in the *Descent*: "Philosophers of the derivative school of morals [i.e., Bentham and James Mill] formerly assumed that the foundation of morality lay in a form of Selfishness; but more recently in the 'Greatest Happiness principle.' According to the view given above, the moral sense is fundamentally identical with the social instincts."[173] Darwin's moral theory simply cannot be reduced to modern kin selection or anchored in utilitarian selfishness. Ruse, though, suggests otherwise.

Ruse believes that Darwin promoted reciprocal altruism and family selection as the foundations for his moral theory, and he interprets both as matters of individual selection—hence ultimately selfish. He denies that Darwin was a group selectionist.[174] Reciprocal altruism could arise, according to Ruse, from so-called epigenetic rules being selected as individual advantages: acting to benefit another on the expectation that one will receive a like benefit in the future. Darwin himself probably thought of reciprocal altruism more in terms of rational calculation and the inheritance of acquired characteristics. But he could have parsed it as resulting from individual selection—he was not explicit about this. Yet as we have seen, this was not Darwin's principal device supporting his theory of moral conscience. To be sure, he did base his notion of moral instinct on the example of selection operative in the social insects—especially bees and ants. But he went beyond what could be interpreted as a version of kin selection to propose the selective advantage of groups that have members who are simply cooperative, altruistic, and self-sacrificing—the kind of group selection that Ruse denies to Darwin. What remains for Ruse is reciprocal altruism, which he thinks is the only grounds left today on which to build an evolutionary theory of morality. His analysis is instructive, for it allows us to dispatch a prevalent objection to an evolutionary ethics, namely, that it is deceptive and really unfit as an explanation of our moral sentiments.

The Darwinian Conception of Objective Morality

In Ruse's estimation, reciprocal altruism arose through individual selection, and in that respect is selfish. He believes that once established in human groups, what began as a selfish motive would not be understood as such. On the surface of consciousness, we would only perceive the feeling or impulse to help another in need. On the phenomenal level, no selfish calculation would be involved. For the moment, I think we can put in abeyance a discussion about the exact mode of the evolutionary development of moral sentiments— whether group selection or reciprocal altruism. Ruse and I agree that such moral sentiments have evolved and have done so under the aegis of natural selection. I want here to investigate what the implications are for our general conception of morality in the Darwinian perspective.

Because Ruse thinks Darwin only endorsed individual selection—and because he himself rejects notions of group selection— he is ready to explain human moral character as the result of a selfish motive, the low motive that Darwin denominated. This is also why he argues that Darwinian naturalism is deceptive, that is, because human beings "are deceived by their genes into thinking that there is a disinterested objective morality binding upon them, which all should obey."[175] The epistemic rules that guide human action, as the result of countless generations of selection on the instinct to reciprocate cooperation, aid, and so on—these rules, according to Ruse, "dispose us to think that certain courses of action are right and certain courses of action are wrong." It is these rules that "give the illusion of objective morality." The rules, however, are entirely subjective and relative to a contingent human history: they are, he says, "idiosyncratic products of the genetic history of the species and as such were shaped by particular regimes of natural selection." We could well imagine, he supposes, "an alien intelligent species evolving rules its members consider highly moral but which are repugnant to human beings, such as cannibalism, incest, the love of darkness and decay, parricide, and the mutual eating of feces." In short, "no abstract moral principles exist outside the particular nature of individual species."[176]

I believe Ruse's conception of morality to be quite antithetic to both Darwin's moral conception and to any notion of morality, tout court. First, it would be unacceptable to Darwin himself, since the master explicitly rejected reciprocal altruism as the primary motive for moral action, since it was based on selfish regard. But more importantly, Ruse's account undercuts any stable conception of morality, since such a conception implies rules of conduct that bind unconditionally.

In order to provide the appropriate analysis, let me consider two meanings for "subjective" and "objective": an ontological meaning and an epistemological meaning. When we refer to a proposition or attitude as subjective, we might simply be claiming that it is held by a subject, a human mind—the ontological meaning. Likewise, when we describe an objective situation, we might mean some event taking place outside the human mind. Epistemically, a subjective belief would be one that is the result of personal idiosyncrasy or prejudice, while an epistemically objective belief—the kind characteristic of scientific propositions—would be one intersubjectively verifiable, capable of confirmation by others acting without bias. I think Ruse slides around between the ontological and epistemic meaning of the terms "objective" and "subjective."

Certainly we deem ethical propositions to be ontologically subjective—they are, after all, held by individual human beings. Need they be epistemically subjective? Certainly not, no more than the logical principle of noncontradiction need be epistemically subjective. Ruse rejects the idea of "morality as a set of objective, eternal verities," and asserts that "no such extrasomatic guides exist."[177] But in these remarks he is arguing, it would appear, that there is no moral code that one could trip over on a hike up the mountain—or maybe, no commandments delivered from the clouds. Certainly, moral codes exist in the minds of individuals; they are ontologically subjective. But this does not mean they are also epistemically subjective. The validity of a given syllogism or even a natural law is objective if intersubjectively verifiable; it can be epistemically objective without being ontologically objective.

It is instructive to compare a moral code with logical principles. Both are ontologically subjective and not ontologically objective. But both can be epistemically objective: we can get intersubjective agreement on both by using standard kinds of tests. In the case of logical principles, we can test whether they preserve truth (e.g., modus ponens, modus tollens, disjunctive syllogism, etc.). Comparably we can test whether a given moral principle preserves other moral principles that are agreed upon. So, for example, if an interlocutor contends that abortion is morally murder but simultaneously admits that the early fetus is not really human, then that would be a test of the principle that abortion is murder, and one that has epistemically objective validity. In the case envisioned, we could confidently claim the interlocutor was holding an epistemically subjective attitude, or, at least, that he or she was confused.

Though human rational capacity is certainly an evolved trait among humans—indeed, if one is totally bereft of that capacity we would be reluctant to regard the creature as human. When Ruse maintains that "morality is rooted in contingent human nature, through and through," he regards that view as disenabling to objective morality. But would he say the same of the rational capacity of human beings—at least in its higher forms, a capacity unique to our species? The fact that only the human brain has evolved to think mathematically does not make mathematical proofs subjective. In both rational argument and moral argument, we might well have a standard (noncontradiction in the one, and altruism in the other), the disposition for which has evolved in our species, but that fact alone does not make the standard or its application epistemically subjective.

This consideration leads to yet another sense in which evolved moral principles can be objective. Among our ancient predecessors, any who met the saber-toothed cat on the path and regarded it simultaneously as dangerous and as gentle—those individuals have not contributed substantially to our gene pool. The principle of noncontradiction has the confirmation of natural selection. Ruse seems to believe that there could be an alien species, in all essentials like us, but one with a moral code that sanctioned cannibalism, incest, and dining on each other's feces. But if that species were intelligent and social, thus like us in that respect, then it would have to be moral like us as well. For what would a social group be like if it were not bound together by ties of cooperation and altruism? Ruse's scenario of social aliens that were nonaltruistic and noncooperative (eating their grandmothers!) would be comparable to a story about an intelligent species that did not observe the principle of noncontradiction. Natural selection would be highly unlikely to have produced such a social or intelligent species.

CONCLUSION

In his analysis of Darwinian morality, Ruse has been misled by thinking Darwin was the intellectual offspring of John Bull. I rather believe he was of mixed cultural parentage. Certainly he inherited a good deal from the traditions of natural theology as well as from the work of British stalwarts like Malthus and Lyell. But there was another, powerful but greatly underappreciated source: namely, the German

Romantic tradition, the work of Humboldt, Goethe, Carus, Bronn, von Baer, Haeckel, yes and even Kant and Schelling. A fair portion of this German tradition, Darwin acquired indirectly from the likes of Richard Owen, especially Owen's reworking of Carus's theory of the archetype. Other features he absorbed from thinkers like William Whewell and many other sources that came to him through translations and personal acquaintance (e.g., Haeckel). Ernst Mayr regarded Darwin as a population thinker, and indeed he was. But populations are groups, and thus it's not surprising that he would be interested in such. In matters of morals, he transcended a British background that emphasized utilitarian judgment, the notion that even our highest moral aspirations stood crippled on the shores of a receding sea of faith. He rather thought his theory brought human beings to a higher level, where the view revealed the grandeur of evolved life.

RESPONSE TO RUSE

To give an account of the accomplishments of a scientist like Darwin requires the exercise of historical imagination, as R. G. Collingwood has termed it. The past must be recreated in the pages of a text, for the past itself does not exist. Strange it is, that as historians we deal with an entity that can only exist as recreation. The past has an inferred existence, since we must use contemporary evidence—books, manuscripts, letters, artifacts—to conjure up an entity that even the historical actors would not quite recognize, given that their own perspectives were confined to a conceptual space much more limited than that of the historian. In 1809, Darwin's father and mother could not have remarked that they had just given birth to the author of the *Origin of Species*. The historian writing today can. The historian, of course, has limited access to the minds of past individuals, but those minds are not completely sealed from the historical gaze. The fears, joys, and ideas of individuals who had once lived are as subject to inferential recreation as much as the public events of the past are. If in our everyday commerce, we could not read the intentions of our fellows, we would have no way of understanding the sounds coming from their mouths, and the case is no different for understanding the behavior of individuals dwelling in the past. We interpret their writings armed with assumptions about their attitudes and intentions. And with Darwin, we have even greater access than to other individuals; he left a huge store of written material that marks out passageways into his creative ideas. Such ideas are often more available to the historian than to Darwin himself, at least during certain times in his career. For instance, in his *Autobiography*, he claimed that he "worked on true Baconian principles, and without any theory collected facts on a wholesale scale."[1] Yet, had he acted as his own historian and investigated his notebooks with a perceptive eye, he would have found entry after entry with the tag "according to my theory."

When reading history, intellectual or social, we also have to make some warranted assumptions about the historian, the methods and modes of analysis, and the questions that guide the recreation. E. H. Carr, the prominent Russian scholar, maintained that as a reader you should "study the historian before you study the facts," since the facts always come tinctured with assumptions, beliefs, and theories that the historian harbors.[2] Ruse and I reconstruct the history of Darwin's thought in light of certain historiographical questions and philosophical interests, which motivate our analyses. This doesn't mean that the facts are simply made up, but rather, that they are put in a certain constellation, in a certain interpretative context that needs to be appreciated so that their implications can be properly assessed. Let me initially indicate the set of assumptions that separate Ruse and me, and then I will consider particular aspects of the history that he portrays.

Ruse and I agree that Darwin helped create "the world in which we now live."[3] That world, especially the scientific world of modern biology, emphasizes the understanding of biological organisms as if they were mechanisms, machines. Ruse quite reflexively employs the comfortable language of contemporary science when characterizing, for instance, Darwin's device of natural selection, referring to it as "the mechanism of natural selection," a locution that Darwin never used in the *Origin of Species*. So the backward slide comes easily, namely, the presumption that Darwin's notion of natural selection depends on, to use Dennett's term, a "mechanistic algorithm." The assumption of mechanism pervades Ruse's analysis. What is needed, I believe, is a careful consideration of the actual language Darwin used in the *Origin of Species*.

A distinct methodological difference also separates my approach to the material from his. Ruse assumes what he should have demonstrated, namely, that the tradition from which Darwin's intellectual ideas stemmed is "quintessentially English."[4] Darwin was an Englishman, no doubt about it, but during his *Beagle* voyage and through his voluminous readings, his ideational nationality enlarged immeasurably. Five years at sea, visiting foreign lands, and carefully studying literature coming from various quarters of Europe (especially the works of Alexander von Humboldt)—all of that will have an effect on one's mental world. During the years following the voyage, Darwin's intellectual horizons continued to expand beyond the shores of Great Britain. This doesn't mean that British literature was absent from his purview: his grandfather's speculations, Lyell's geology, Malthus's theory of population—this and much more certainly occupied his attention. But equally important were the studies coming from France (e.g., Lamarck, Cuvier, Geoffroy Saint-Hilaire, de Candolle) and Germany (e.g., Humboldt, Goethe, Carus, von Baer). Moreover, a fair amount of British literature conveyed conceptions originally derived from the continent, especially from Romantic Germany: Richard Owen's ideas about the archetype and homology had their origin in the work of Kant, Goethe, Schelling, and Carus; Owen's depiction of the theory of recapitulation, which he rejected but Darwin adopted, derived from Lorenz Oken and Carus. William Whewell's challenge to explain the progressive advancement of species by natural law had its roots in Kant's rejection of any tincture of theology in natural science. So the horizon of Darwin's thought extended far beyond the white cliffs of Dover.

Before even considering Darwin's intellectual development, Ruse goes on at great length depicting what he takes to be the social and conceptual environment of Great Britain ("Britain before Darwin"), as if one could simply assume that this environment must have decisively molded the naturalist's conceptions. As an impressionable youth, Ruse himself came under the spell of the Cambridge historian Robert Young. Young's Marxist assumption that the base of British industrialism and economy determined the superstructure of science left on the neophyte a lasting impression. We should take our cue from perceptive biologists themselves, who have come to realize that animal and plant organisms may exist in an environment that's radically different from the apparent surroundings. To take a trivial example, the layman might assume that a bat swooping down to catch insects is led by their color or smell, while the zoologist, who has first assessed the capacities of the animal itself, recognizes the importance of reverberating sound. Ruse begins with what he takes as Darwin's social and conceptual environment and then moves inwardly, presumptively to pin down the young researcher's developing ideas. I rather begin in an attempt to assess the ideas that Darwin actually expressed, and then move outwardly to determine their sources. Of course, in both of our efforts there is a reciprocal movement back and forth, but it is a matter of different emphasis.

As Carr suggested, study the historian first. Ruse and I have intellectual predilections that focus on different philosophical traditions, which inform our respective approaches to the historical material. He's born into a tradition that he outlines in respect to Darwin, British empiricism, with David Hume as the standard bearer; though his lingering social constructionist—rather Marxist—attitude might have given Hume terminal dyspepsia. My own tendencies are toward the neo-Kantian tradition, which includes German idealism. Neither of us simply adopts these orientations uncritically, but they do reflect, I think, the kinds of sensitivities and assumptions we make. I'm sure this is not exactly a revelation to an attentive reader.

In what follows, I will discuss five pivotal topics about which Ruse and I radically differ. We would both agree that these topics are central to understanding Darwin's accomplishment: the use of metaphor in the *Origin*; the role of teleology; the conception of progress; individual selection versus group selection; and the evolution of human morality.

THE LANGUAGE OF METAPHOR

In the *Origin*, Darwin deployed metaphors almost reflexively. He remarked, for instance, on the struggle that goes in the guise of "the face of nature bright with gladness."[5] He drew the ghastly consequences of this struggle, in which "the face of Nature may be compared to a yielding surface, with ten thousand sharp wedges packed close together and driven inwards by incessant blows."[6] The unset-

tling, mixed trope of this latter analogy is an index of Darwin's im-
pulsive habit to think metaphorically. The metaphor of "the face
of nature" works implicitly at two levels. It first suggests that the
struggle in nature usually goes on without our recognition and that,
second, beneath that irenic veil, competitive struggle eliminates
many varieties. But Darwin was also reflective about his use of metaphor, when, for example, he stated: "I should premise that I use the
term Struggle for Existence in a large and metaphorical sense."[7] He
wished to apply the term not only to dogs struggling over a piece of
meat, but to a plant at the edge of a desert struggling for moisture
or mistletoe struggling with other plants to have its berries dissemi-
nated by birds. In this use of metaphor, Darwin explicitly indicates
how flexible his central principle of struggle is, and how a number of
phenomena are unified under its rubric.

Most theories in science have an associated model that articulates
the features of the theory. Models are, of course, metaphors; they de-
pict a more familiar structure in order to articulate a less familiar
one. In my historical reconstruction, I tried to show how Darwin's
imaginative notion of a powerful and moral intelligence served as a
model for natural selection in the essays and in the *Origin*. Ruse re-
joins that this "anthropomorphism" was not essential to Darwin's
argument, and he quotes Darwin as later suggesting that this usage
was "merely" metaphorical.[8] But this rejoinder misses the point:
while not to be taken literally—nature per se is not intelligent—the
metaphor of an intelligent mind yet structures nature quite differently than if the metaphor were that of a machine. It allowed Darwin
to understand how nature could operate over vast periods of time
and select the smallest advantages, could see into the fabric of organisms, and ultimately could act morally for the benefit of "each individual." No mechanism of Darwin's acquaintance could perform in
this "intelligent" and "moral" way. By contrast, Huxley believed the
course of evolution would not be slow and gradual, but intermittent
and jumpy; but then he thought of nature in mechanical terms, quite
unlike his friend. Moreover, in the development of his theory, Darwin
explicitly conceived the actions of natural selection as a secondary

cause, with the divinity as the primary cause, the ultimate source of a "designed" law. In this way, nature took on *virtual* intelligence and moral concern.

Darwin's doubts about the existence of God and divine governance grew over time, so that he finally admitted he was in a muddle over the matter. Ruse focuses on this later state of Darwin's religious trajectory, but simply does not appreciate the way the naturalist's religious convictions controlled the development of his theory before 1859; these convictions left their ineradicable imprint on his conception of nature and remained ever a part of the core of the theory through all of the editions of the *Origin of Species*.

Another telling instance of the way metaphor controlled the development of Darwin's conceptions concerns his principle of divergence, a principle he thought the "keystone" to the *Origin of Species*.[9] The principle of divergence aims to explain why species fall into morphological clusters that can be represented by a branching tree, with wider or narrower gaps between the branches. This clustering allows the naturalist to apply the taxonomic categories: varieties, species, genera, families, and so on. Darwin believed he had neglected this problem before the 1850s, and only solved it while riding in his carriage through the Kent countryside.[10]

The path to his formulation of the principle of divergence is winding and complex. He gave it fairly clear expression in his *Big Species Book*: "in the long run, more descendants from a common parent will survive, the more widely they become diversified in habits, constitution & structure so as to fill as many places as possible in the polity of nature, the extreme varieties & the extreme species will have a better chance of surviving or escaping extinction, than the intermediate & less modified varieties or species."[11] The principle, at first reading, seems bizarre. Why should extremes have an advantage over less extreme forms? Aren't extremes of size, appearance, and habit just as likely to be disadvantageous? This looks almost like a principle of hopeful monsters. Here again, Darwin's model, his metaphor, gets absorbed into the theory. The model was based on his own recent experience as a pigeon breeder.

To explain the function of the principle in nature, Darwin first described how divergence worked in the domestic sphere:

A fancier is struck by a pigeon having a slightly shorter beak; another fancier is struck by a pigeon having a rather longer beak; and on the acknowledged principle that "fanciers do not and will not admire a medium standard, but like extremes," they both go on . . . choosing and breeding from birds with longer and longer beaks, or with shorter and shorter beaks. . . . Here, then, we see in man's productions the action of what may be called the principle of divergence.[12]

Darwin simply assumed that nature had a preference for extremes comparable to that of a pigeon breeder—an assumption quite consistent with the model of an intelligently functioning process. My point is that one has to pay close attention to the language Darwin uses, and not quickly amalgamate his ideas to our modern theory.

When Ruse refers to Darwin's principle of divergence, he parses it in terms of the *division of labor*.[13] He recognizes that Darwin also used this metaphor, but does not bother to examine the language more precisely, simply happy that it has a British provenance— even if Darwin got it more proximately from the Frenchman Milne-Edwards. In order to understand how Darwin applied the notion of the division of labor one must closely investigate the metaphors he used in the application. So let me explore Darwin's metaphorically infused argument.

Milne-Edwards referred to the differentiation of parts in a single organism: an organism that used a stomach for both digestion and breathing lacked the advantage of an organism that divided the labor into a stomach for digestion and lungs for respiration.[14] Since about 1854, Darwin thought of Milne-Edwards's notion of the division of physiological labor as a mark of higher development.[15] So when he considered the divergence of groups on their way to becoming morphologically separate species, a process he thought progressive, he unhesitatingly applied the metaphor of division of labor to that process. And here is where the trouble began—or rather, continued. As in the case of the pigeon-breeder metaphor, Darwin was led astray when utilizing the trope of the division of labor. Metaphor got the better of him and led him into advancing a principle—the principle of divergence—that really made no sense and was superfluous. Why is that the case?

If two groups of closely related organisms attempt to occupy the same environment, then one of two scenarios must occur: either one drives the other to extinction or one, because of competition, adapts to slightly different ecological conditions and thus diverges in morphological structure from the other. In both cases, it is simply a matter of natural selection operating on small morphological differences in particular environments, and no principle of the division of labor or of divergence is required. In the latter instance, if natural selection slowly works on individuals of a species that have wound up in slightly different environments, those individuals will gradually diverge in their traits in so far as they become adapted to those different environments; no other principle, not one based on extremes or on the division of labor, is necessary.[16] This is simply a further example of the power of metaphor to structure a theory often beyond what the scientist may have reflectively found acceptable. In Darwin's case, the application of the metaphor of the pigeon breeder (and that of the division of labor) measures the depth of his commitment to the assumption that nature acts with something like intentional goals.

What I believe to be defective about Ruse's analysis of Darwin's theory is the assumption that metaphors are only decorative and can be safely ignored in the construction of a scientific theory. I believe they do real work, even if that work is sometimes misguided, as I'll further specify in a moment.

TELEOLOGY

Huxley initially thought that teleology—that is, the doctrine of final causes—had been banished from biology with the advent of Darwin's theory:

> The Teleology which supposes that the eye such as we see it in man or one of the higher *Vertebrata*, was made with the precise structure which it exhibits, for the purpose of enabling the animal which possesses it to see, has undoubtedly received its death blow.[17]

This cannot be quite right, since eyes have been naturally selected to produce vision. Organisms with a modicum of sight or with keen sight have the advantage over organisms lacking this trait or having it in weaker measure. So, many creatures have been designed with eyes by natural selection for the "purpose of enabling the animal which possesses [them] to see." Of course, Huxley understood this. What he denied was that a benevolent intelligence stepped into nature and provided the organism with the advantage of sight. Ruse agrees that organisms express design, though he qualifies this attribution by saying that for Darwin it was only "as if" organisms were designed by a superior intelligence. Moreover, he denies that Darwin conceived any directionality to evolutionary development. Of course, these presumptions are commonplaces of modern biology: nature has no goals; and we are not the apple of nature's eye. But these common attitudes of today may not be perfectly compatible with the theory constructed by Darwin yesterday. Since there are several other meanings for *final cause*, it would be well to identify them and decide which are tolerated by Darwin's theory and which are not.

Huxley himself proposed another meaning for final cause that would be consistent with Darwin's conception. He indicated that because Darwin believed in the absolute determinism of nature by law, it could be said that the "existing world lay, potentially, in the [original] cosmic vapour."[18] But this is not a terribly interesting meaning for final directionality in nature. One might identify another meaning of teleology that stems from Kant—and ultimately Aristotle— that would be compatible with Darwin's theory, namely, that all of the parts of an organism are tightly related to other parts as reciprocally means and ends: the heart pumps blood in order that the liver can function; and the liver purifies the blood in order that the heart can function. Darwin could accept this notion of teleology, though he would have added that there was a looseness about the relation-ship of parts, so that small changes might be introduced without dis-rupting the whole. (Cuvier and Lyell had objected to Lamarckian evo-lutionary theory because any newly evolved attribute would tend to undo the whole organism, so tightly connected, they believed, were a creature's traits.) A related meaning for final cause recognized by Kant (and Aristotle) is that the purpose of one member of a species is to produce another member of that species—organisms are de-signed for reproduction, the engine of evolution; machines, at least of Darwin's acquaintance, were not capable of replication. He cer-tainly could endorse the purposive, or goal-directed, character of or-ganisms in these constrained senses and still stay true to his theory.

In these several meanings for teleological causality, an anteced-ent state is explained in terms of a subsequent state: eyes have the purpose of sight; the heart has the purpose of oxygenating the other parts of the body; organisms have the purpose of reproduction. The consequent state—the purpose or final cause—makes intelligible the antecedent state. But if that were the only claim, there would likely be no objection of a good Darwinian to this sort of teleology. After all, we better understand a structure if we know what it typi-cally produces, what its function might be. But can we say that the antecedent structure (the eye, for example) came into existence due to the subsequent state or function? We can, if we understand that productive relationship in terms of natural selection always work-ing to improve the functioning of structures in a given environment. Yet, in the logic of Darwin's theory, this latter productive relationship ultimately amounts to this: a variation has occurred because of some efficient cause (i.e., some antecedent cause, say, the environmen-tal impact on the sexual organs of the parent), and if that variation works better in the organism's environment—or is at least neutral—then that variation would be retained and likely would be further built upon. This kind of operation of selection is a bit like Michel-angelo's removing the objectionable parts of the Carrera marble to leave the beneficial parts; and so the statue of David emerges.

There is yet another meaning of teleology, also expressed by Kant. He argued that the design features of an organism cannot be explained mechanically, but must be assumed to have been the product of a plan, an idea; and such ideas, he further argued, could only be produced by an intellect. So the naturalist, in giving an account of the teleological features of organisms, must ultimately assume they have arisen because of a plan formulated by a powerful intelligence, an *intellectus archetypus* in Kant's terms.[19] He cautioned, however, that this assumption could not be explicative in true natural science (*Naturwissenschaft*), but could only serve as a regulative assumption, a heuristic that allowed the naturalist to explore further what might be the mechanistic causes acting on the organism. Kant was yet convinced that a complete mechanistic understanding of organisms was impossible, which is why he believed that biology could not be a science (*Naturwissenschaft*) but only a discipline or doctrine (*Naturlehre*). As I indicated in my essay, William Whewell comparably recognized that organisms displayed purposive structures and that, over vast periods of time, they progressively advance toward a goal, human beings. But he did not think any natural law could explain the purposive structure of creatures or their progressive development over time. One must turn to theology for the proper account. William Paley, by contrast, contended that the purposive structure of organisms could be explained by natural law. The deity, he proposed, did not intervene in nature or bend natural laws to create species; rather He formulated natural law so that the structure of species could be entirely explained by such law. The laws themselves, Paley contended, also required an explanation. They were due to the wise provisions of the Creator. Thus the laws were secondary causes having God as the primary cause.[20]

I believe that Darwin threaded his way between the positions of Whewell and Paley. He maintained that natural law, especially natural selection, could explain what Whewell thought could not be so explained, namely, the progressive development of species, having the final goal of human beings; and, like Paley, he proposed that these laws were "designed." They were produced by an intelligent mind governing the universe.[21] Given these assumptions, Darwin could advance arguments that were distinctively teleological (e.g., "man is one great object, for which the world was brought into present state"),[22] but that did not fall outside the pale of natural science. Darwin explicitly referred to these as "designed laws." We should not be surprised, then, when we see in the *Origin* Darwin claiming that natural selection worked for the good of each creature or that nature's grandeur consisted in the production of human beings as moral agents.

Darwin did not hesitate: the laws of nature were "designed," and thus gave a certain shape to creatures and their evolutionary trajectory; and those laws were the product of a superior intelligence. Yet Ruse contends that this is all "as if." But there is not the slightest indication in the notebooks, essays, or the *Origin* that Darwin was arguing as a Kantian, offering an *als ob* argument. To argue *als ob* would not be an adequate response to Whewell's challenge of producing a law that could explain design. Darwin was not crossing the boundaries of his discipline, to argue as a philosopher (Ruse's assumption) or as a theologian (Sober's assumption). He was using a proper teleological argument, subtle though it be, to explain the goal of nature — human beings. There are no grounds for trying to escape the plain language of Darwin's text.

EVOLUTIONARY DEVELOPMENT AS PROGRESSIVE

Early in his scholarly career, Ruse helped establish the orthodox view about the nonprogressive character of Darwin's theory. In his *Darwinian Revolution* (1979), he detailed the progressivist conceptions of the early evolutionists Jean-Baptiste de Lamarck, Robert Chambers, and Herbert Spencer, and then concluded that Darwin's theory was radically different on this score. Where Lamarck saw "a progression to man," Darwin "did not see progress to man."[23] Stephen Jay Gould likewise argued that "an explicit denial of innate progression is the most characteristic feature separating Darwin's theory of natural selection from other nineteenth century evolutionary theories. Natural selection speaks only of adaptation to local environments, not of directed trends or inherent improvement."[24] Natural selection, after all, only leads to local improvement; when the environment changes those improvements would be lost. Gould thought ideas of progress were tied to European imperialism and racism, and he was sure that Darwin—and contemporary biologists who adopted his theory—could not be guilty of these sins. Peter Bowler filed a concurring opinion: "Darwin's mechanism [of natural selection] challenged the most fundamental values of the Victorian era, by making natural development an essentially haphazard and undirectional process."[25] The assumption that Darwin's theory was nonprogressivist has even led to the strange notion that he did not use the word "evolution" in his works because that term implied progressive development. But, of course, "evolution" appears as the last word in the *Origin of Species*: "from so simple a beginning endless forms most beautiful and most wonderful have been, and are being, evolved."[26] And the term is used liberally in the *Descent of Man* and in later editions of the *Origin*.[27]

Yet upon due reflection over the years, neither Ruse nor Gould could deny that the text of Darwin's *Origin* was replete with the language of progress, and both came to propose an amendment to their earlier claims: namely, that Darwin's ideas about progress were not part of his science, rather they were part of a cultural doctrine that he impressed on his science. Darwin, in Ruse's estimation, was committed to an ideology of a

> God who works only through unbroken law. But, whether this is legitimate or not, it was a vision that Darwin was imposing on the organic world, not a deduction made by reasoning from the evidence.[28]

Again, this conclusion supposes that philosophy is a distinct enterprise from science—a twentieth-century view imposed on a child of the nineteenth century.

In our debate, I'm perplexed about Ruse's current construal of the text. On the one hand, he agrees that Darwin "believed in progress, with the system topped out by humans. . . . This was all part and parcel of the deist world picture he embraced."[29] Yet he also thinks Darwin never accepted nature as manifesting a series of stages, "all going up to the climax, humankind. This was not Darwin's picture of things."[30] On the surface, at least, these seem like perfectly contradictory interpretations. Perhaps, however, Ruse is merely contending that the progressive development of nature was not automatic, not somehow generated from the internal structure of organisms (where Lamarck located the progressive impulse), but was a consequence of the relentless actions of natural selection. Darwin certainly did believe that progressive development was the general tendency of selection: "And as natural selection works solely by and for the good of each being, all corporeal and mental endowments will tend to progress towards perfection." In the third edition of the *Origin* (1861) he refers to the progressive action of selection as "inevitable."[31]

If species development is progressive, what is it progressing toward? Indeed, the notion of progress requires there be a goal *toward which*, a terminus *ad quem*. From the very beginning of his theorizing about species change, Darwin conceived the trajectory of nature as aiming toward human beings ("Man is one great object, for which the world was brought into present state").[32] Indeed, before he hit upon Milne-Edwards's criterion of the division of labor as a standard of progress, Darwin simply used human beings as that standard, as he remarked in a note dated March 1845: "What is the highest form of any class? Not that which has undergone most change. For changes may reduce organization:—generally, however, that which has undergone most changes & which approaches nearest to man."[33] Ruse observes that in the third edition of the *Origin*, Darwin expressly used Milne-Edwards's criterion for calibrating progressive development. In that same edition, he equated the division of labor with the simpler criterion he had already formulated in 1845: "the degree of intellect and an approach in structure to man clearly come into play [when judging the highness of other vertebrates]."[34] Thus human beings established the goal of nature and the criterion by which other creatures were evaluated for progressive development.

This teleological goal also explained various other features of natural history, for example, the great age of the earth. For if biological nature moved in minute steps keeping pace with geological change, and if man manifested the very complex traits of high intelligence and moral capacity, then long periods would be required to prepare for the advent of human beings. As Darwin succinctly put it: "Progressive development gives final cause for enormous peri-

ods anterior to Man."[35] Darwin's "one, long argument" would ultimately make little sense if those "designed laws" he spoke of did not
lead to the "highest object we are capable of conceiving," namely,
man with his moral sentiments. If we finally agree that Darwin was
a nineteenth-century mind—though laying the ground for the new
scientific world that would shortly come cycling into view—then it
should not be surprising that he was a progressivist and one who believed human beings to be the highest accomplishment of nature—
a vision also quite consonant with that of his Romantic contemporaries in Germany.

INDIVIDUAL VERSUS GROUP SELECTION

Both Ruse and I regard Darwin as more than a two-trick pony: evolutionary descent and natural selection. He was immensely creative
throughout his life, while retaining a fund of ideas that, like geological structures, only very slowly changed. Yet on the issue of individual
selection versus group selection, Ruse seems to think that Darwin's
ideas stagnated. He's invested in making Darwin an individual selectionist, since he seems to think that's what a quintessential Englishman ought to be. Any trucking around with group ideas would be to
slide into the abyss of German Romantic thought. Put in these terms,
the issue may seem of slight consequence, but it is not. It has special
relevance for the very meaning of natural selection and for Darwin's
moral theory. Before I address Darwin's attitude about group selection, we should be clear about what he understood to be the logic
of selection: as applied to the individual, as applied to a group of related individuals (e.g., the social insects), and as applied to a group
of largely unrelated individuals.

In chapter 3 of the *Origin*, Darwin defined natural selection in
terms of the struggle for existence in this way:

Owing to this struggle for life, any variation, however slight and
from whatever cause preceding, if it be in any degree profitable to an
individual of any species, in its infinitely complex relations to other
organic beings and to external nature, will tend to the preservation of

that individual, and will generally be inherited by its offspring. . . . I have called this principle, by which each slight variation, if useful is preserved, by the term Natural Selection.[36]

There are three features of this definition to emphasize: 1) selection is a process that preserves individuals in the struggle for existence because of their useful traits; 2) individuals, being preserved to re-productive age, will tend to pass those traits to offspring; and 3) those traits will be preserved in the society (and spread) because their car-riers each generation have been preserved.

In the case of community or family selection (e.g., in the social in-sects), useful traits would be predicated of the hive as a unit.[37] Thus a hive of bees that produced, by chance, several aggressive defenders (i.e., soldiers) would have the advantage over other hives that had fewer or no aggressive defenders. One should note two things here: that having more aggressive members cannot be a trait of individu-als, only of a collective; and that the aggression of solider bees offers them as individuals no advantage, really a disadvantage because it puts them at peril. Since the advantageous trait is a property of the whole community, the community would have greater opportunity to survive and thus reproduce (i.e., send out queens to establish new hives); and those new hives would also tend to have the trait of pro-ducing aggressive individuals. In the case of social insect communi-ties, the individuals of those communities are all related as members of an extended family. Hence, to act cooperatively or altruistically would be, in a sense Darwin dimly perceived, to act for the benefit of one's own hereditary structure. Thus Ruse is ready to parse commu-nity selection as a kind of individual selection.

Ruse and I will understand group selection to involve communi-ties or collections of largely unrelated individuals. In analogy with individual and family selection, any advantageous trait would have to be predicated of the whole group, giving the group a survival ad-vantage. The largest groups at issue in Darwin's discussions would be either the variety (a subspecies) or the entire species. In any case of group selection—including species selection—the advantageous property must belong to the group as a whole.

Now, let me address the question of Darwin's attitude about group selection from two approaches, both of which Ruse considers: the effort to explain equal sex ratios in different species, and his relationship with Wallace.

Could natural selection produce equal sex ratios in those species that exhibit such equality (the human species, for instance)? Ruse argues that Darwin tried to handle this problem in the *Descent of Man* only by application of individual selection: "Darwin's discussion made it clear that he never thought to solve the problem through group selection."[38] I can only conclude Ruse and I must be reading entirely different books.

In the first edition of the *Descent* (1871), Darwin recognized two conditions for natural selection to produce equal ratios: an advantage for equality of sex ratios and for reduced fertility. Say a species had more males than females; those extra males, Darwin contended, would be unproductive and thus, he maintained, a *disadvantage to the species*. Those species that had greater equality would be more productive of progeny. "But our supposed species would by this process be rendered, as just remarked, more productive; and this would in many cases be far from an advantage" (because the environment might be unable to sustain a larger population, forcing severe competition among species members). Hence, "a simultaneous decrease in the total number of the offspring would be beneficial, or even necessary, for the existence of the species."[39] There can be no doubt that Darwin thought that equal sex ratios with lower numbers of offspring would be to the direct advantage of the species, not the individual. This can only be group selection, where the group is the species. We should be clear about this. Males, in Darwin's example, that are non-productive (because of fewer females) would not be at a disadvantage in the struggle for existence—they already exist and would tend to reach reproductive age, though without consequence. Producing fewer or more progeny is not an advantage to the individual; it's only a consequence of reaching reproductive age. Moreover, Darwin expressly claimed in these passages that the greater productivity would be an advantage, not to the individual but only to the species.

In the second edition of the *Descent* (1874), Darwin recognized, as

Ruse points out, that he couldn't quite make the argument for equal ratios work. He confessed: "I formerly thought that when a tendency to produce the two sexes in equal numbers was *advantageous to the species*, it would follow from natural selection," but he perceived the problem was too complicated to resolve.[40] (What Darwin may have sensed as the problem was that the struggle, in this scenario, would have to go on among species, which seems unlikely.) The point of my objection to Ruse is simply that Darwin tried a group selection account—in the form of species selection—and saw no conceptual obstacle per se to such an effort. Darwin had more than one tune in his natural selection repertoire.

What about his relationship to Wallace on the issue of group selection? Scholars have recognized that Wallace and Darwin, though they came independently to similar theories, differed in some distinctive ways. For instance, Darwin used the production of domestic varieties as a model for the production of incipient species in nature, while Wallace initially rejected domestic production as a model since he thought artificial varieties would naturally revert to the parent type.[41] The two naturalists would also come to differ about sexual selection, Darwin holding that the males of many bird species, for example, attained their gaudy colors because of the choice of the females, while Wallace believed the default state, as it were, had both sexes highly colored but that selection drove the females to drab colors for reasons of protective camouflage. And as Ruse and I have indicated, Wallace could come to find the ultimate source of man's high intelligence and moral capacity in the operations of the spirit world. We part, however, in our understanding of the two naturalists' conception of natural selection. Ruse thinks that Wallace was more disposed to group selection, while Darwin remained an individual selectionist.

Ruse attributes Wallace's inclination toward group selection to his socialism, while Darwin remained Adam Smith's man. Indeed, Ruse suggests that Darwin's insistence on individual selection is a mark of the deep English character of his theory. I think Ruse wrong about both Wallace and Darwin. Ruse depicts three aspects of Wallace's and Darwin's theories, made over the extent of their friendship, to in-

dicate the difference between the two on the question of group selection: the way Wallace characterized struggle for existence (i.e., natural selection) in 1858, the dispute the friends had over the origins of
hybrid sterility, and their conceptions of the origins of morality.

In 1858, Lyell and Hooker arranged to have the essay that Wallace
sent to Darwin about species change and excerpts from Darwin's
1844 essay (along with a letter to Asa Gray) published in the Linnaean
Society journal. The publication would allow both to be credited with
the discovery of evolution by natural selection, though Darwin would
be seen to have the priority. In a recently published article, which is
epitomized in his current essay, Ruse supposes that Wallace's 1858
essay characterized selection as essentially acting on the varietal
group instead of the individuals of that group. This idea has become
orthodox with many historians of biology.[42] The passage Ruse quotes
is one in which Wallace is explaining how a variety of a species might
displace other varieties, including the parent species, in the struggle
for existence:

> It is evident that, of all the individuals composing the species, those
> forming the least numerous and most feebly organized variety would
> suffer first, and, were the pressure severe, must soon become extinct.
> The same causes continuing in action, the parent species would next
> suffer, would gradually diminish in numbers, and with a recurrence
> of similar unfavourable conditions might also become extinct. The
> superior variety would then alone remain on a return to favourable
> circumstances would rapidly increase in numbers and occupy the
> place of the extinct species and variety.[43]

Granted that Wallace's expression of the struggle for existence is
a bit vague from our perspective, but then he was not writing with
our special question in mind. Yet it is clear enough from this quotation that Wallace assumed that the *individuals* of a variety, in the
imagined circumstance, would suffer selection against, since their
traits were unorganized and there were fewer such individuals.
These same causes—operating against the *individuals* of the parent
stock—would reduce their numbers, driving them to extinction. This

analysis is no different from Darwin's own in the second chapter of the *Origin*, where he maintains that "if a variety were to flourish so as to exceed in numbers the parent species, it would then rank as the species . . . or it might come to supplant and exterminate the parent species."[44] Darwin, here, is simply classifying individuals as having certain advantages that would allow them to defeat another group of individuals that were classified as the parent species. In both Wallace's account and Darwin's, the individuals were described according to the general properties they possessed—their varietal properties—which properties gave each individual either an advantage or disadvantage. There is no ascription of properties to the class per se, as opposed to the individuals in the class. In Wallace's book *Darwinism* (1889), there is no doubt that he represented his notion of selection as the same as Darwin's. He provided what he thought a perfect example of Darwinian selection at work: the wingless condition of many insects—flies, moths, and beetles—on the Kerguelen Island, an extremely windy location. The insects must have reached the island because of wings, but could survive on the island only if they lost their wings. A given variety of moth, for instance, might be characterized as having the advantageous property of being wingless, but only in virtue of its individual members being wingless—the class, per se, would not have the property.[45] Wallace in this example and in his 1858 paper thus understands selection operating directly on individuals, even if the advantageous or disadvantageous property is common to all the members of the group. The property in question is not a feature of the group per se (as it would be were it, say, a complex social structure).

Ruse thinks that Wallace proposed a group selection explanation in his dispute with Darwin over the origins of intervarietal infertility, while Darwin regarded such infertility as an accidental by-product and not any trait selected for. To be sure, Darwin could not conceive any advantage to the individual for the trait of diminished fertility (except in very special circumstances). Wallace focused on the rise of infertility between a parent species and its daughter variety, which itself was becoming a new species in its own right. In Ruse's interpretation, Wallace assumed that selection worked "for the benefit of the parent species," and thus "wipes out the inadequate mixed off-spring." But this is not what Wallace argued in the letter to which Ruse refers. In the letter, Wallace only maintained that he could show "a considerable amount of sterility would be advantageous to a variety" (not to the parent species).[46] In the body of the letter, Wallace did not explain how sterility could be an advantage, but in a several-page enclosure he did provide an explanation. The explanation is rather convoluted, but it does not rely on group selection. Wallace imagined a scenario in which two varieties of a species became separated in a region, with each adapting to the particular local environment; and at the edges of their local regions, hybrids would be formed. Assume, further, that for contingent reasons, the pure forms had a greater disinclination to cross with one another than the hybrids. Now if selective pressures would generally increase, the pure forms, being better adapted to their regions, would have the advantage over the hybrids, and would push the hybrids to extinction while establishing the pure forms with the greater disinclination to cross — and the process would be continued.[47] Wallace's account may be contrived — and he returned to have another go at the explanation later in the century, though with no more success[48] — but nothing in his account speaks of group selection.

These correctives concerning Wallace's position on individual versus group selection provide further evidence of the liabilities of moving from the external social environment (i.e., Wallace adopting the socialism of Robert Owen) to fixing a biological theory, rather than initially determining the precise character of the biological theory and then moving to discover what social causes—if any—might be at work. I have used these correctives also to prepare the way for the more important consideration, namely, the similar accounts of morality offered by both Wallace and Darwin, which do rely on group selection.

In my essay, I pointed out many passages in the later editions of the *Origin* and in the *Descent* where Darwin did not scruple to invoke a group selection account without any concern for the relatedness of individuals in the group (or anticipating any modern inclusive fitness explanations). I also tried to show that even when Darwin appeared to make an appeal to community selection—a community interlaced with related members—his proposal was not a version of a kind of Hamiltonian kin selection. I will say a bit more about this below. In what follows, I will approach Darwin's theory of morality in a somewhat different way from what I presented in my essay.

THE EVOLUTION OF MORALITY

To understand Darwin's account of the evolution of morality two general considerations have to be brought to the fore: his solution to the problem of the wonderful instincts of the social insects, and his use of Wallace's early explanation of human evolution, the one his friend offered before the conversion to spiritualism. And in the deeper background lies Darwin's experiences during the *Beagle* voyage.

I gave a brief account in my essay of Darwin's discovery of family (or community) selection in the case of the social insects. Ruse is right, this became a model, in the *Descent of Man*, for understanding the evolution of the moral instinct in human beings. It is, however, uncertain when Darwin perceived his solution to the social insect problem as relevant to the evolution of moral behavior. Much earlier, when he first turned to consider human morality at the beginning of September 1838, just before he read Malthus, he was completely in Paley's camp. He had attempted a biological version of Paley's rule of utility, which stated: "Whatever is expedient is right. But then it must be expedient on the whole, at the long run, in all its effects collateral and remote, as well as in those which are immediate and direct."[49] Darwin suggested that those useful and expedient habits that were necessary to preserve animal groups, allowing them to propagate and protect their young (habits of sociality, friendship, etc.), were what we had come to call morally good. The continued practice of such behaviors, over the generations, would cement them into the inherited structure of the organism. Paley assumed that pleasure in exercise would move individuals to follow the rule of utility. Darwin became sensitive to this selfish motive in Paley's proposal after he read Sir James Mackintosh's critique of Paley. Darwin essentially adopted Mackintosh's position, finding confirmation in a view of innate behavior that was not based in selfishness.[50] In this early period, Darwin supposed that the cause instilling the behavior in humans was the inheritance of acquired characters. Only with his solution to the problem of the social insects did he have another possible causal ground for moral behavior.

We must not be misled by Darwin's model of the social insects, however. He used the model to explain how behaviors that do not provide an advantage to the actor but to the recipient might yet evolve under natural selection. The worker bee who acts to keep the nest clean expends energy in the service of its community; the soldier bee, going even farther, may expend its life for the community. The peculiarity of social insect communities is that, generally, the workers, drones, and queen are related as sisters, brothers, or mother. Small human tribal communities in early times were likely bound by blood relationships, as Ruse contends; and so altruistic behavior might be expended on another individual, who might be a sibling, parent, grandparent, cousin, or aunt or uncle. Of course, only from the modern perspective of the "selfish gene," might such altruism be interpreted as ultimately selfish behavior. But there is the further analytic point that I've already mentioned but needs to be reiterated. In the human case, Darwin typically imagines tribal societies composed of some family members but including others who are not so related. The society that has individuals who are more cooperative, faithful, and patriotic will have the advantage over those having fewer such individuals. But having more such altruistic individuals can only be a property of the group; the advantage accrues to the group as a whole, "the individuals composing the community being at the same time indirectly benefited."[51] Ruse is simply mistaken about Darwin's reliance on individual selection in the *Descent of Man*.

The additional stimulus to Darwin's construction of the moral theory that occupies two chapters in the *Descent of Man* was Wallace's paper before the London Anthropological Society in 1864. I cited that paper in my essay and showed how certain passages are almost identical to those Darwin penned in the *Descent*. Darwin thought Wallace's essay novel, and he admired it greatly. Here's another passage from Wallace that certainly won Darwin's approval. It expresses the deep humanity both researchers shared:

> By his [man's] superior sympathetic and moral feelings, he becomes fitted for the social state; he ceases to plunder the weak and helpless of his tribe; he shares the game which he has caught with less active or less fortunate hunters, or exchanges it for weapons which even the sick or the deformed can fashion; he saves the sick and wounded from death; and thus the power which leads to the rigid destruction of all animals who cannot in every respect help themselves, is prevented from acting on him. This power is "natural selection."[52]

As is obvious, Wallace's version of the application of natural selection to human groups was not confined to bestowing benefits on blood relatives, but on all human beings within the society, an expansion upon which Darwin also insisted. Here is a comparable passage in the *Descent* in which Darwin likewise extends the moral attitude far beyond the narrow social group, foretelling a time when human "sympathies became more tender and widely diffused, so as to extend to the men of all races, to the imbecile, the maimed, and other useless members of society, and finally to the lower animals,— so would the standard of his morality rise higher and higher."[53] Such an extension could not occur through individual selection.

One should not forget that Darwin's own moral feelings extended far beyond his loving father, brother, and sisters. When he experienced the institution of slavery in South America during his voyage, he expressed the same morally indignant rage as had his guide, Alexander von Humboldt, and he so expressed it in his *Journal of the Voyage of the Beagle*:

> On the 19th of August we finally left the shores of Brazil. I thank God, I shall never again visit a slave-country. To this day, if I hear a distant scream, it recalls with painful vividness my feelings, when passing a house near Pernambuco, I heard the most pitiable moans, and could not but suspect that some poor slave was being tortured, yet knew that I was as powerless as a child even to remonstrate. . . . It is argued

that self-interest will prevent excessive cruelty; as if self-interest protected our domestic animals, which are far less likely than degraded slaves, to stir up the rage of their savage masters. It is an argument long since protested against with noble feeling, and strikingly exemplified, by the ever illustrious Humboldt.[54]

Darwin composed this passage during the time he had just finished his 1844 essay; it is from the second edition of his *Journal of the Voyage of the Beagle*, written when he would have been deep into his theory.

Darwin claimed that his version of morality escaped "the reproach of laying the foundation of the most noble part of our nature in the base principle of selfishness."[55] The altruistic instincts stilled by natural selection operated without any hedonistic calculations. They didn't ultimately derive from what Darwin called a "lower motive," resulting from reciprocal altruism or from praise and blame. The higher social animals, he believed, differed from humans in this respect: baboons, for instance, developed great canine teeth through male combat and for self-defense, though they also might be used to defend the tribe. But none of these animals acted "solely for the good of the community." In the case of human beings, however, the moral and intellectual faculties "have been chiefly, or even exclusively, gained for the benefit of the community; the individuals composing the community being at the same time indirectly benefited."[56] This is simply the result of natural selection being applied to the group, that expanding circle of the moral community that both Darwin and Wallace envisioned.

CONCLUSION

A reader might find it strange that two individuals who seem to know the facts of a situation can so vehemently disagree about the understanding of those facts, as Michael Ruse and I do. One salient difficulty is that we are exploring questions that Darwin himself did not often pose—for example, about the entrenched force of metaphor or about the intricacies of group selection. Occasionally he did seem to reflect about these issues; but usually they require from the historian a probing below the vaguely troubled surface of Darwin's thought. This does not mean the facts upon which we base interpretations cannot be objective, but only interpretation all of the way down, as it were. Facts are not like Silly Putty that can be stretched in one way and then unrecognizably in the other. But each of us—and the reader—has to struggle to find that layer upon which there is inter-subjective agreement and then to lay out that solid material carefully to construct an argument upon it. False steps can be made at any time in the process, but our mutually critical endeavor ought to provide enough material and enough cogent argument for the reader to make a decision. At least, this is what we hope.

REPLY TO RICHARDS

It is remarkable how Darwin recognizes among beasts and plants his English society with its division of labour, competition, opening up of new markets, "inventions," and the Malthusian "struggle for existence."

KARL MARX, letter to Friedrich Engels[1]

Humboldt's conception of nature as a creative force and a repository of moral and aesthetic values would lie at the foundation of all of Darwin's later work on species, and especially the human species.

ROBERT J. RICHARDS, *Was Hitler a Darwinian?*[2]

As noted in the introduction, there is a huge amount of totally uncontested material pertinent to Charles Darwin and his achievements. We know about his schooling, his travels, his family, his friends and his enemies and detractors, the times when he made his major discoveries and when he wrote them up and made them public, and much, much more. Darwin at this level is not one of those people shrouded in mystery and forever beyond our grasp. This said, reading Robert J. Richards's essay and then rereading my essay, I am struck by how completely different are our two visions. Let there be no mistake, to use a hackneyed term, Richards and I are in different paradigms.

I set my story against the Darwin-Wedgwood family background deeply immersed in the Industrial Revolution. This means competition, it means Progress, and above all it means machines. Against this—and I see this as a complement not a contradiction—I stress Darwin's early Christian training, at home, at school, and then at the

University of Cambridge. Truly in his *Autobiography* did he acknowledge his debt to Archdeacon Paley, the author of the leading textbooks on British theology generally, writing of them that "careful study of these works, without attempting to learn any part by rote, was the only part of the Academical Course which, as I then felt and as I still believe, was of the least use to me in the education of my mind."[3] An entirely natural consequence of this upbringing was that for many years, up to and including the *Origin*—later the belief declined—Darwin assumed that an all-powerful deity designed and created everything. However, note precisely what this means. Part of the assumption about the deity is the Christian belief, especially the evangelical Christian belief—the influence of Darwin's childhood—that there is a strict separation between Creator and created. The belief that the material world, including the organic life to be found in this world, has no meaning or value in and of itself. God, the Supreme Industrialist, makes the machines—the telescope-like eye, for instance—and any meaning and with it value comes through, and only through, God.

Darwin may or may not have known much of the philosophy of Descartes—it is hard to imagine that when he studied Locke's *Essay concerning Human Understanding* at Cambridge the name never occurred although there are no mentions in the *Correspondence* and only fleeting references in such works (read by Darwin) as Malthus's *Essay* and Lyell's *Principles*. My claim is simply that Darwin, the British mechanist—reinforced by his Christian training—bought in completely to the Cartesian view of the world as being *res extensa* (material or extended substance) without mind or thought (except inasmuch as these can be explained materially) and extended this (as Descartes would not) to humankind and its thinking. Things in themselves just are. Even vaunted, human, unique features like free will are illusory, under the mechanical picture as interpreted by Darwin—"one doubts existence of free will every action determined by heredetary constitution."[4] This means, and here I am entirely in line with orthodox interpretations of the Scientific Revolution, God moves from being all-important to being irrelevant.[5] I doubt Darwin ever became a full-blooded metaphysical naturalist, but—combined with his theological worries about the Christian God—his methodological naturalism was pushing him that way.

Richards starts with mind and feeling. His Darwin goes back to Plato and perhaps even beyond to Parmenides. The world is not dead. It is alive and feeling, filled if not with intelligence then at least with some kind of life force or spirit. It is a world with value in itself. For this reason, ultimately all is one, there are connections throughout nature. To use Plato's language, everything comes from the Form of the Good. We humans are part of this, even though as it happens we have a special place in this vision, namely, at the top. Plato believed in a designer (the Demiurge of the *Timaeus*) and for much of his life Darwin believed in a designing Creator, but this is not so much unimportant as beside the point. What is to the point is that the world itself is more than dead matter.

> And I have felt
> A presence that disturbs me with the joy
> Of elevated thoughts; a sense sublime
> Of something far more deeply interfused,
> Whose dwelling is the light of setting suns,
> And the round ocean and the living air,
> And the blue sky, and in the mind of man;
> A motion and a spirit, that impels
> All thinking things, all objects of all thought,
> And rolls through all things.[6]

I do not say that Richards's Darwin is an explicit Platonist. I do say that Richards's Darwin stood in this tradition—a tradition that at the end of the eighteenth century, especially in Germany in its literature and its science, took the form of Romanticism. Charles Darwin was at one with this and his science; particularly his theory of evolution through natural selection manifests and confirms this fully.

Do I exaggerate our differences? You might think so. Talk of "paradigms" suggests that Richards and I are in some real sense in different worlds. We may use the same language, but we refer to different things. I am not really interested here in continuing debate about one of the slipperiest terms in the English language. Let me rather pick out a couple of items on which we disagree to show how very deep are our disagreements, so much so that we are in a sense talking past each other. I'll start with a topic where it is I who am trying to push the argument, because the points I am making are absolutely fundamental to my thinking. Richards is not so much playing catch-up, but his thinking is perhaps more a consequence of his overall position. Then I will reverse the pattern, taking up a topic absolutely fundamental to Richards's thinking and where I am more fitting my thinking into my overall position. Having done this, I will then turn to discussion of why Richards and I are so far apart.

LEVELS OF SELECTION

If, in the past forty years, there is one topic that has been discussed—overdiscussed—by historians and philosophers of evolutionary biology, it is the question of the unit of selection.[7] This is no great surprise, because it is a topic much discussed by practicing evolutionary biologists and we historians and philosophers have been pulled in. I don't think anyone now wants to use the past to show that we today are necessarily better. It is more a matter of finding heroes from the past who can cloak our present thinking with authority and respectability. I have certainly played this game, although because one has an agenda this does not mean that one is necessarily wrong. Be that as it may, in today's language (which I noted was not that of Darwin and his contemporaries) I see Darwin as an individual selectionist, where the organism is the main focus of selection. Richards sees Darwin as a group selectionist, where the group can be the focus of selection, and is indeed so in certain very important cases, most notably when it comes to the production of the human moral sense. Complicating this discussion, as readers will know, is the fact that we both allow Darwin's acceptance of a form of family selection. We differ however on the meaning and significance of this. Where we do not differ is in seeing what may seem like an arcane and technical point as something that divides us fundamentally and points to our different philosophies and visions.

For me, it is absolutely essential that natural selection be something acting at and basically only at the individual-organism level. Darwin didn't know about genes, so he could not be a "selfish gene" supporter, but he was as close to that as it is possible to be. Note that being in favor of selfish genes does not spell the end of cooperation. Organisms can be "altruistic," but only if it rebounds to their biological advantage. Reciprocal altruism is one way that this can happen. Another way, much favored today, is through kin selection, where helping relatives funnels back to one's own improved biological fitness.[8] Obviously Darwin didn't have and couldn't have our notion of kin selection, especially when today's biologists start showing how the hymenopteran reproductive system can yield odd results (like nonreproductive female workers). But he did grasp the notion of relatedness and for him this was the way in which he could stay true to his vision of individual selection. For Darwin, in some sense the family was a super-organism, with the individual relatives parts of the whole and not entities in their own right. I thus see family selection as an extension of individual selection. Richards, whose Darwin sees connections and wholes throughout nature, quite comfortably goes the other way, taking family selection as the lowest level of a full-blooded group selection, where the unit of action can be a group (like a species) of nonrelated members. For Richards, it is hardly something to be argued. It is a consequence of the Romantic vision. There would be something odd if Darwin's explanation of human morality did not rely on selection between (unrelated) groups. To insist on relatedness in some way hints of self-interest and, say what you like, this is not really compatible with true moral thought and behavior. One does not have to be a Kantian to see that real giving never counts the cost.

Why am I so insistent on individual selection? I think I have made it clear that it is here perhaps more than anywhere I make the connection between Darwin and his British background and heritage, especially the industrial-Whig side in which his family were embedded. Adam Smith is all-important. Say it again: "It is not from the benevolence of the butcher, the brewer, or the baker that we expect our dinner, but from their regard to their own interest." There are some (not

Richards) who have questioned this link, arguing that there is no evidence that Darwin ever read the *Wealth of Nations*. (I note that he certainly read something by Smith, but more likely the *Theory of Moral Sentiments*.) I find this objection unconvincing. Whether he read the book or not, I cannot believe that a child of the Darwin-Wedgwood family would never have heard of Smith's economic philosophy.[9] This was a family where moral and philosophical issues were discussed—Darwin not only read Sir James Mackintosh but this writer on philosophical topics was a family visitor—and most certainly socioeconomic topics would have been a constant subject in brother Erasmus's London set that Darwin joined for a period after the *Beagle* voyage. Don't forget, the money came from a family working according to Smith's precepts. In any case, we do know that in the spring of 1840 Darwin read a historical survey of political economy that had a full and favorable exposition of Smith's thinking.

> Dr. Smith has . . . shown, in opposition to the commonly received opinions of the merchants, politicians, and statesmen of his time, . . . that it is in every case sound policy, to leave individuals to pursue their own interest in their own way: that, in prosecuting branches of industry advantageous to themselves, they necessarily prosecute such as are, at the same time, advantageous to the public; and that every regulation intended to force industry into particular channels, or to determine the species of commercial intercourse be carried on between different parts of the same country, or between distant and independent countries, is impolitic and pernicious—injurious to the rights of individuals—and adverse to the progress of *real* opulence and lasting prosperity.[10]

I will not labor this point further. I have argued that not only does Darwin adopt an individual-selection stance but that, even if this were at first done subconsciously, his dispute with Alfred Russel Wallace in the 1860s made him both consciously aware of the issues and determined to defend the position he had taken from the first.[11] Wallace was not a Romantic. He was a socialist. Different philosophies but they run parallel in stressing the group over the individual. Expectedly if we look at the paper Wallace sent to Darwin and that was published in 1858 by the Linnaean Society, we find that already there Wallace is far more inclined to a group position than is Darwin. For Wallace, individual struggle seems more of a purifying process within the group. If you are second rate, you are going to get wiped out. Change is a group phenomenon. One variety is going to do better than another. The very title of the essay flags you to this: "On the tendency of varieties to depart indefinitely from the original type." More than just this. The change doesn't seem to come about because one variety goes to war with another variety and wins. At least not directly. It is all a matter of change of circumstances and of one variety doing better in the new circumstances than do other varieties. This is very different from Darwin's picture where certainly change can occur because of change of circumstances—a new predator, for instance—but change can also occur internal to the group, for instance, when one form uses a little less food than the others. It is no different from one mill owner getting an edge over his competitors because he learns to use his resources a little more efficiently and thus improves his profit margins.

EMBRYOLOGY

Turn now to embryology. Darwin himself stressed how very important this was to him, writing to his friend Joseph Hooker just after the *Origin* was published and moaning: "Embryology is my pet bit in my book, & confound my friends not one has noticed this to me.—"[12] Some fifteen years later he confirmed this sentiment, writing in his *Autobiography*: "Hardly any point gave me so much satisfaction when I was at work on the *Origin*, as the explanation of the wide difference in many classes between the embryo and the adult animal, and of the close resemblance of the embryos within the same class."[13] Both Richards and I pick up on this point, and each spends considerable time discussing the significance of embryology in Darwin's thinking about evolution. And yet, how very far apart we find ourselves, not the least because for Richards embryology is where it all starts and ends whereas for me it is something to be fitted into an existent whole.

Richards is the world expert on the Romantics' view of organic development, and I, unhesitatingly, am his student.[14] One starts with the individual organism, a vertebrate let us say, beginning with a very primitive life form and developing by the time it is an adult into a fairly sophisticated, more advanced life form. There is a kind of inherent direction, a teleology, to this growth, with a sort of inner momentum taking one through the stages from beginning to end. Then one sees (and here the Platonism can be discerned) that the developing parts of the organism are not separate, but generally all formed from the same basic units. The backbone of the vertebrate is a perfect example, with the same bone form repeated again and again. But there are more complex and sophisticated repetitions. To take the most famous example, as first pointed out by Goethe, the vertebrate skull seems itself to be modified units of the backbone. This is within individual organisms. One finds also that there are shared patterns between organisms of different species, cows, say, and horses. They may end at different destinations, but they have common starting points. The embryos of very different adults are alike. One finds also that there are connections between the stages. There is a correspondence between the development of the individual and the range of organisms extant today. Some of the latter are relatively primitive (fish, for instance), and some fairly complex and advanced (higher mammals, for instance). What is remarkable is that in the development of the more complex organism one sees the unity of all life, because such development entails going through the adult forms of the less complex. At an early stage of development, mammalian embryos are fishlike to the extent that one can even say that they are fish. Capping this and again showing that all is one we find that as the history of life is unfurled, the earlier organisms were more primitive. So in fact, we have not only a parallel between the development of the individual and the range of living beings, but another parallel between these and the history of life forms. All of this is not necessarily evolutionary, but it can be interpreted as such. The late nineteenth-century

German evolutionist Ernst Haeckel was famous for his dictum that "ontogeny," the development of the individual organism, recapitulates "phylogeny," the development or history of the group. And as a finale to all of this, we have a picture that makes humans important above all other organisms. We are the most complex, the end point of the progressive ride upward, the apotheosis of the Romantic picture.

Richards sees Darwin as a total enthusiast for this understanding of life and its driving forces. It is no exaggeration to say that Richards sees Darwin's biology, including his evolutionary biology, as built on this Romantic framework. Richards's Darwin sees connections between individual body parts (what Richard Owen called "serial homology"), and analogously his Darwin sees connections between the embryos of very different species (what Richard Owen called "general homology"). "Darwin drew support for his view by appeal to Goethe's theories of the ideal plant and the vertebral theory of the skull—thus aligning himself with these Romantic ideas." Richards's Darwin sees upward progression, as organisms go through the life forms of more primitive organisms. "Darwin joined the idea of the archetype with another Romantic theory, namely, the recapitulation hypothesis—that the ontogenetic development of the embryo mirrored the phylogenetic development of the species."[15] And at the top comes our species. "If man is *one* great object for which the world was brought into present state,—a fact few will dispute, . . . & if my theory be true then the formation of sexes rigidly necessary.—"[16]

I go in a very different direction, and start by noticing how for me, for all that I accept the great importance of embryology for Darwin, I see it as something that fits into the theory rather than a starting point for the whole enterprise. I see embryology as part of the picture of selection working its way out, and I see Darwin's *vera causa* strategy, particularly the analogy with artificial selection, as crucial to Darwin's supportive argumentation. In other words, it is British empiricism that is crucial to my understanding of Darwin's embryology. Of course, I see Darwin starting with embryological similarities between organisms different as adults. Obviously he learned this from his mentors—men like Grant and Henslow—and as obviously Germanic science is in the background. Grasping this fact was crucial to Darwin's systematic work in his lengthy study of barnacles. I also see Darwin accepting that in individual growth we are often going to see stages that remind one of adult forms of other, less complex, organisms. Mammals do have a fishy stage. But at once I put this in a selection context. There is nothing inevitable about all of this. Organisms tend to add on to what they have. It is these add-ons that get the full blast of selection. A mammal is a mammal because it has adaptations that enable it to live its mammalian existence that is different from—I very much suspect Darwin would say, superior to—the existence of fish. Why don't mammals lose their fishy stages? Basically because selection doesn't care. Embryonic mammals are protected from the environment, they do just fine as they are, and so they remain unchanged. Darwin proves this with his reference (in the passage I quote in my main essay) to the breeders of dogs and horses.

> The case, however, is different when an animal during any part of its embryonic career is active, and has to provide for itself. The period of activity may come on earlier or later in life; but whenever it comes on, the adaptation of the larva to its conditions of life is just as perfect and as beautiful as in the adult animal. From such special adaptations, the similarity of the larvæ or active embryos of allied animals is sometimes much obscured; and cases could be given of the larvæ of two species, or of two groups of species, differing quite as much, or even more, from each other than do their adult parents.[17]

Stemming directly from this position is the fact that for Darwin any recapitulation would never be of adult forms but always of embryos. Evolution is a matter of adding on to the embryonic forms. Here his thinking is in line with the great embryologist Karl Ernst von Baer (whom Darwin misidentifies as Agassiz in the first edition of the *Origin*) who saw development as stemming out from original types without inevitable repetition. "Fundamentally the embryo of a higher animal form never resembles the adult of another animal form, but only its embryo."[18] So really it is incorrect to say of Darwin that he subscribed to Haeckel's biogenetic law: Ontogeny recapitulates phylogeny. And most importantly there is the question of humans. I agree that Darwin saw evolution as progressive with humans at the pinnacle. What I do not see is humans as necessarily the end point or even the only end point. How could they be given Darwin's arms-race-type solution to the problem of progress? In any case, who comes out on top is for God to decide. Let me quote yet again a passage quoted by Richards and repeated by me just above. This time I will quote the passage in full, including a line dropped by Richards. "If man is *one* great object for which the world was brought into present state, — a fact few will dispute, (although, that it was the sole object, I will dispute, when I hear from the geologist the history, from the astronomer that the moon probably is uninhabited) & if my theory be true then the formation of sexes rigidly necessary. — "[19] On the one hand, this really does show how Richards and I inhabit different worlds. I think the bracketed line is part of the original thought; he thinks it added as an afterthought at most qualifying a basic position. On the other hand, I agree that my extended version does not refute the idea that God created humans through an upward force, nor does it prove that humans like everything else came through the unguided process of natural selection. But I do think the extended version fits more comfortably with the notion that everything is subject to the same natural laws, no exceptions even for us.

THE ROMANTIC INFLUENCE

Why are Richards and I so far apart? What is behind our rival visions? You know my reasons: the family and the education. What is Richards's equivalent? Obviously he stresses different influences. Above all Richards sees the lasting influence of the Romantics on the young Darwin. He thinks that this influence completely colored the thinking of Darwin and that this set the program that reached its highest points in the *Origin* and the *Descent*. In response, the one thing I am not about to do is deny the influence of Romanticism on the young Darwin. He grew up at a time when it was flourishing as never before or since, and he was drawn to areas like biology where the philosophy, if one might so call it without prejudice, struck deeply and profitably. There is no need to hypothesize. We have massive documentary evidence of the input. For instance, near the beginning of this response I quoted Wordsworth's *Tintern Abbey*. We know that precisely around the time when he was developing his theory of evolution, Darwin read and enjoyed the works of this poet and his fellows. In his *Autobiography*, of the early years after the *Beagle* voyage, he wrote, "About this time I took much delight in Wordsworth's and Coleridge's poetry," and he repeated that sentiment later in the work.[20] Coleridge I need hardly say was even more immersed in Romanticism than Wordsworth, and was given to copying out and publishing without attribution large portions of Schelling's philosophical writings.

You might think that in itself reading poetry is hardly that relevant to the work of a scientist. I am a huge fan of "Tartan Noir," the school of crime fiction with roots going back to James Hogg and Robert Louis Stevenson; but I doubt my somewhat narcissistic fondness for brilliant, incorruptible policemen, with major Oedipal problems, a string of broken relationships, and dietary practices that are low even by Scottish standards, greatly influences my thinking about matters Darwinian, philosophical or historical. In Darwin's case I suspect we can make a stronger connection, at least as far as style is concerned. For instance, in the evolution notebooks we find him thinking about the past in near poetical terms and referencing Wordsworth. "I a geologist have illdefined notion of land covered with ocean, former animals, slow force cracking surface &c truly poetical. (V. Wordsworth about science being sufficiently habitual to become poetical)."[21] The pertinent passage in Wordsworth, from the preface to his *Lyric Ballads*, reads: "If the time should ever come when what is now called Science, thus familiarized to men, shall be ready to put on, as it were, a form of flesh and blood, the Poet will lend his divine spirit to aid the transfiguration, and will welcome the Being thus produced, as a dear and genuine inmate of the household of man."[22]

At least as or more important than the poets was the morphologist Richard Owen. Although Owen later became a bitter enemy of Darwin and his group, in the years after the *Beagle* voyage, Darwin and Owen were in close contact, brilliant young scientists rising up in their field, and Owen was far more sophisticated and learned about biology than Darwin and taught the young traveler a lot. And what Owen was teaching Darwin was a great deal of Romantic biology—the hypotheses and findings of what were known as the Nature Philosophers (or Scientists), the *Naturphilosophen*. Even Darwin admitted this, if grudgingly. "I often saw Owen, whilst living in London, and admired him greatly, but was never able to understand his character and never became intimate with him."[23] There is no doubt that Owen was always touchy, and he would have resented the young naturalist—rich, handsome, worldly, the pet of the Cambridge circle and friends (Whewell, Henslow, Sedgwick, Lyell). But the evolutionary notebooks speak of the influence and in particular they speak of the influence of one who much admired German biology, even as his reliance on the Cambridge circle (who were very suspicious of what they took to be the pantheism of German Romanticism) made Owen careful to keep his fullest speculations privately to himself. To Darwin perhaps he could be more candid. "Every successive animal is branching upwards, different types of organization improving as Owen says, simplest coming in and most perfect and others occasionally dying out."[24] "Mr Owen suggested to me, that the production of monsters (which Hunter says owe their origin to very early stage) & which follow certain laws according to species, present an analogy to production of species."[25] "Curious paper by M. Serres on Molluscous animals representing foetuses of vertebrata, &c. 1837, p. 370 Owen says nonsense."[26] "All classes of *Acrita* exhibit lowest stages of animal organization, '& are analogous to the earliest conditions of the higher classes during which the changes of the ovum or embryo succeeded each other with the greatest rapidity'—so we find species each class successively present modifications typical of succeeding classes & likewise those much higher in scale. So Owen actually believes in this view!!!"[27] And there is much more. Darwin's use of German biology in the barnacle work surely owed much to Owen's teaching. The same is true of the evolutionary discussions, in the preliminary essays and in the *Origin*.

I do not however see the full-blooded *Naturphilosoph* program being transferred across to Darwin, more than we saw in the last section. Darwin is interested but certainly does not appear as a convert to Owen's full-blooded Germanic biology, and that seems to me just about right and expected. Nor, expectedly, would Darwin have learned only German biology from Owen. Darwin's worrying about the social insects shows just how important was the inclusion of behavior in the scheme of things. This is not a particularly Germanic insight. It is there in Kirby and Spence. It is strongly reinforced by Owen. "Owen says 'the necessity of combining observation of the living habits of animals, with anatomical & zoological research, in order to establish entirely their place in nature, as well as fully to understand their oeconomy, is now universally admitted.'"[28] Later in life, Owen became a kind of idealistic evolutionist, and he may well have been inclined that way when Darwin first met him in the late 1830s. But by nature and politically (he needed the favor of powerful people for financial support) Owen could not afford to be too forward-thinking or too revealing of his hand. Don't forget either that balancing Owen would have been Whewell who shows himself in his *History* (which Darwin read twice in 1838) to be cautious about German biology and Sedgwick who would have been quite hostile. It is true that in the early evolutionary writings, Darwin picks up on and cites favorably the vertebrate theory of the skull. Why wouldn't he? It all seems grist to the evolutionary mill. But when in 1858, Huxley attacked the theory (in his Croonian Lecture to the Royal Society), Darwin (in the *Origin*) dropped it sharply.[29] It just wasn't that crucial to him. Natural selection would be quite happy to make use of materials already at hand, but if push comes to shove, it can make something new.

ALEXANDER VON HUMBOLDT

What about the elephant in the room? I refer to the naturalist and travel writer Friedrich Wilhelm Heinrich Alexander von Humboldt (1769–1859). Let's start off with two incontestable facts. The first is that Humboldt was the Romantics' Romantic. He was in total agreement with Goethe and others who shared that philosophy. He saw the world as a whole, as one. About to set out on the travels to South

America (1799–1804) that made him famous, Humboldt wrote in a parting letter: "I shall try to find out how the forces of nature interact upon one another and how the geographic environment influences plant and animal life. In other words, I must find out about the unity of nature."[30] And again and again his reaction to the living world showed that for him it was an aesthetic and moral experience, as he subsumed his very being in the environment around him. "This influence of the physical on the moral world—this mysterious reaction of the sensuous on the ideal, gives to the study of nature, when considered from a higher point of view, a peculiar charm which has not hitherto been sufficiently recognized."[31] This is not a man in the Cartesian world of *res extensa*.

The second fact is that Humboldt was a massive influence on and inspiration for the young Charles Darwin. In his *Autobiography*, Darwin wrote truly: "During my last year at Cambridge I read with care and profound interest Humboldt's *Personal Narrative*. This work and Sir J. Herschel's *Introduction to the Study of Natural Philosophy* stirred up in me a burning zeal to add even the most humble contribution to the noble structure of Natural Science. No one or a dozen other books influenced me nearly so much as these two. I copied out from Humboldt long passages about Teneriffe."[32] It was Humboldt who inspired Darwin to travel, as of course he did on the *Beagle* voyage (1831–36). Darwin's letters bear out fully his enthusiasm for Humboldt, and his account of the *Beagle* voyage, published first as part of the official account of the voyage and then separately as (in the second edition) the *Voyage of the Beagle*, the work that made Darwin a household name, is full of praise for Humboldt. Talking of the scenery he encountered: "As the force of impressions generally depends on preconceived ideas, I may add, that all mine were taken from the vivid descriptions in the Personal Narrative of Humboldt, which far exceed in merit any thing I have read on the subject."[33] Not that Darwin was slow to try his own hand at the job. Here he is as he steps out into the New World.

The day has past delightfully. Delight itself, however, is a weak term to express the feelings of a naturalist who, for the first time, has been wandering by himself in a Brazilian forest. Among the multitude of striking objects, the general luxuriance of the vegetation bears away the victory. The elegance of the grasses, the novelty of the parasitical plants, the beauty of the flowers, the glossy green of the foliage, all tend to this end. A most paradoxical mixture of sound and silence pervades the shady parts of the wood. The noise from the insects is so loud, that it may be heard even in a vessel anchored several hundred yards from the shore; yet within the recesses of the forest a universal silence appears to reign. To a person fond of natural history, such a day as this, brings with it a deeper pleasure than he ever can hope again to experience.[34]

And in concluding reflection Darwin returns to the theme. "Among the scenes which are deeply impressed on my mind, none exceed in sublimity the primeval forests undefaced by the hand of man; whether those of Brazil, where the powers of Life are predominant, or those of Tierra del Fuego, where Death and Decay prevail. Both are temples filled with the varied productions of the God of Nature:—no one can stand in these solitudes unmoved, and not feel that there is more in man than the mere breath of his body."[35] So Humboldtian did Darwin get at times, that one of his sisters reproved him gently.

I have been reading with the greatest interest your journal & I found it very entertaining & interesting, your writing at the time gives such reality to your descriptions & brings every little incident before one with a force that no after account could do. I am very doubtful whether it is not *pert* in me to criticize, using merely my own judgment, for no one else of the family have yet read this last part—but I *will* say just what I think—I mean as to your style. I thought in the first part (of this last journal) that you had, probably from reading so much of Humboldt, got his phraseology & occasionly made use of the kind of flowery french expressions which he uses, instead of your own simple straight forward & far more agreeable style. I have no doubt you have without perceiving it got to embody your ideas in his poetical language & from his being a foreigner it does not sound unnatural in him—[36]

And yet! The question is not Darwin as a travel writer, but Darwin as a scientist. Darwin as an evolutionist who changes our world. Here I think we must tread carefully. A youthful enthusiasm does not necessarily translate into a mature creative thinker. In later years, Darwin himself rather recognized this fact. To be fair, Darwin did see, in some of Humboldt's speculations, more than a hint of Lyell's theory of climate. "I may remark that the general horizontal uplifting which I have proved has & *is now* raising upwards the greater part of S. America & as it would appear likewise of N. America, would of necessity be compensated by an equal subsidence in some other part of the world. — Does not the great extent of the Northern & Southern Pacifick include this corresponding Area?"[37] Darwin noted that this kind of thinking can be found in Humboldt, who remarks that "the epoch of the sinking down of Western Asia coincides with the elevation of the platforms, of Iran, of central Asia, of the Himalaya, of Kuen Lun, of Thian Chan, and of all the ancient systems of Mountains, directed from East to West."[38] And Darwin did (at the end of his life) praise Humboldt for making a science of biogeography. "I have always looked at him as, in fact, the founder of the geographical distribution of organisms."[39]

However, even in 1845, by which time Darwin with reason was thinking of himself as a mature geologist, he was (in writing to Lyell) critical of Humboldt's flowery philosophical style (in *Cosmos*) — "the semi-metaphsico-poetico-descriptions in the first part are barely intelligible" — and even more so of the geology — "I grieve to find Humboldt an adorer of Von Buch, with his classification of volcanos, craters of Elevation &c &c & carbonic-acid gas atmosphere."[40] And again late in life Darwin was generally rather cool toward Humboldt as a scientist.

> I believe that you are fully right in calling Humboldt the greatest scientific traveller who ever lived. I have lately read two or three volumes again. His Geology is funny stuff; but that merely means that he was not in advance of his age. I should say he was wonderful, more for his near approach to omniscience than for originality. Whether or not his position as a scientific man is as eminent as we think, you might truly

call him the parent of a grand progeny of scientific travellers, who, taken together, have done much for science.[41]

Note that even if Humboldt did speculate in ways that would support Lyell's climate theory, he did not have the theory itself and was not observing constantly with an eye to supporting the theory—something that Darwin did throughout the voyage.

What about evolution and natural selection? Humboldt was a Romantic and, as noted, the unity of life was fundamental.

> I have described the sensations, which the cow-tree awakens in the mind of the traveller at the first view. In examining the physical properties of animal and vegetable products, science displays them as closely linked together; but it strips them of what is marvellous, and perhaps also of a part of their charms, of what excited our astonishment. Nothing appears isolated; the chemical principles, that were believed to be peculiar to animals, are found in plants; a common chain links together all organic nature.[42]

Darwin certainly picked up on some of this: "Again there is beauty in rhythm & symmetry, of forms—the beauty of some as Norfolk Isd fir shows this, or sea weed, &c &c—this gives beauty to a single tree,—& the leaves of the foreground either owe their beauty to absolute forms or to the repetition of similar forms as in angular leaves,—(this Rhythmical beauty is shown by Humboldt from occurrence in Mexican & Graecian to be single cause) this symmetry & rhythm applies to the view as a whole.—"[43] Although there are certainly sources closer to home who stressed these sorts of themes. Given his sexual appetites—five children with a first wife, two with a mistress (the governess), almost certainly another child with another (married) woman, and then seven children with another wife (to whom he was writing poetry until her first husband conveniently died)—there is little surprise that grandfather Erasmus Darwin put matters in a predictable context: "A Grecian temple may give us the pleasurable idea of sublimity"[44] and "when any object of vision is presented to us, which by its waving or spiral lines bears any similitude to the form of the female bosom, whether it be found in a landscape with soft gradations of rising and descending surface, or in the form of some antique vases . . . , we feel a general glow of delight."[45]

As Richards has argued convincingly, thoughts of transmutation were part and parcel of the Romantic vision, and Humboldt is certainly at one with this kind of thinking. Speculating on the distributions of organisms, he commented: "This phenomenon is one of the most curious in the history of organic forms. I say the history; for in vain would reason forbid man to form hypotheses on the origin of things; he still goes on puzzling himself with insoluble problems relating to the distribution of beings."[46] Darwin was obviously aware of all of this and in thinking of and reading and rereading Humboldt, he was using the older man's works as both inspiration and foil. Somewhat amusingly there is in Humboldt's account of his travels support for my claim that when, in his discussion of the evolution of morality, Darwin refers to tribes he is referring to what he takes to be groups of interrelated individuals (families). "The Guaicas, whom I measured, were in general from four feet seven inches to four feet eight inches high (ancient measure of France). We were assured, that the whole tribe were of this extreme littleness; but we must not forget, that what is called a tribe constitutes, properly speaking, but one family."[47] Perhaps more significantly, after the Malthusian breakthrough there is a remarkable scribbled comment in the margins of Humboldt's discussion of jaguars, their rarity, and their different races.

> To show how animals prey on each other—what a "positive" check. Think of death only in Terrestrial vertebrates—smaller carnivora— Hawks. What hourly carnage in the magnificently calm picture of Tropical forests. Let him from some pinnacle [view] one of these Tropical [forests] how peaceful and full of life. Probably two or three hundred thousand jaguars in S America. What slaughter & as many pumas.[48]

All of this said, though, it does not add up to massive input. As Darwin himself recognized, in another comment scribbled in the margins of Humboldt's narrative, "Nothing respect to species theory."[49]

Even though Darwin was to praise Humboldt as a biogeographer, note how in the passage quoted above, he writes of "insoluble problems relating to the distribution of beings." From the first, Darwin was working in a Lyellian mode, using the distributions to ferret out the past, and then as he developed his own theory, using the distributions to ferret out the evolutionary past.

But still could one not say that metaphysically Humboldt's influence was there through and through Darwin's thinking, especially in the first edition of the *Origin*? Natural selection is not then the survival of the fittest. It is natural *selection*. Selection implies a selector, a conscious mind or at least some kind of world spirit. Right at the heart of his theory Darwin shows his debt to Romanticism, because his force of change is something that necessarily implies a living world in some sense. *Res extensa* has no place for this kind of thinking.

> How fleeting are the wishes and efforts of man! how short his time! and consequently how poor will his products be, compared with those accumulated by nature during whole geological periods. Can we wonder, then, that nature's productions should be far "truer" in character than man's productions; that they should be infinitely better adapted to the most complex conditions of life, and should plainly bear the stamp of far higher workmanship?
>
> It may be said that natural selection is daily and hourly scrutinising, throughout the world, every variation, even the slightest; rejecting that which is bad, preserving and adding up all that is good; silently and insensibly working, whenever and wherever opportunity offers, at the improvement of each organic being in relation to its organic and inorganic conditions of life.[50]

This is not dead matter going through its paces.

Let me agree! Right up through and including the *Origin* we have something more than molecules in motion. And that something more includes something living in some way, that infuses, combines and makes sense of the whole of reality, especially organic reality. The question is what are we talking about, and here we see starkly the difference between my vision and that of Richards. I see the world of British natural theology. I see the other great influence on Darwin— as given in the quotation above from his *Autobiography*—is Herschel, who says that all force is will force and ultimately this is the force of the deity. I see Lyell, with his deistic vision of the world, designed by God to keep turning over and over again like the Newcomen engine, and I see Lyell's disciple Charles Darwin scampering all over South America trying to prove Lyell right. I see Archdeacon Paley stressing the design-like nature of organisms, and Darwin taking this—final cause—to be the defining nature of what it is to be living and think-ing up natural selection to speak explicitly to this issue. I see human-kind as central to this vision—we are uniquely made "in the image of God."

But I also see God as the ultimate living being that gives force and direction and value. It is not nature in itself. To quote a contempo-rary evangelical, the head of biology at Wheaton College (the alma mater of Billy Graham): "Scripture provides a logical value system. It establishes that the whole creation in general, and every part of it in particular, has a value given to it by God. This does not mean that the creation is inherently good or that it has the right to exist on its own merits, independent of God. Its goodness is derived from its Creator and so is a kind of 'grace' goodness, freely given in love, not grudgingly merited by right."[51] It is within this framework that natu-ral selection fits in perfectly. Indeed, if you want to say that at the earliest stages, around 1840, Darwin was even thinking of selection anthropomorphically, or theologically in terms of the Creator, I will not deny you absolutely.

> Who, seeing how plants vary in garden, what blind foolish man has done in a few years, will deny an all-seeing being in thousands of years could effect (if the Creator chose to do so), either by his own direct foresight or by intermediate means,—which will represent [?] the creator of this universe.[52]

Anthropomorphic or not, however, it is God who is doing the work and not nature. And this now sets me on the path that takes me ever farther from Richards. Notice how, from the very beginning, although God stands behind everything, Darwin is getting God out of the immediate business of creating and getting everything done through general law.

> That all the organisms of this world have been produced on a scheme is certain from their general affinities; and if this scheme can be shown to be the same with that which would result from allied organic beings descending from common stocks, it becomes highly improbable that they have been separately created by individual acts of the will of a Creator. For as well might it be said that, although the planets move in courses conformably to the law of gravity, yet we ought to attribute the course of each planet to the individual act of the will of the Creator. It is in every case more conformable with what we know of the government of this earth, that the Creator should have imposed only general laws.[53]

And this of course intensifies and continues right down through the *Origin*. In the letter to Gray Darwin makes it quite clear that he does think that God is ultimately behind everything and doing the creating, but it is always done through the medium of unbroken law. In this respect, as Darwin makes clear in a letter written to Lyell just after the *Origin*, biology is no different from astronomy. "I doubt whether I use term Natural Selection more as a Person, than writers use Attraction of Gravity as governing the movement of Planets &c. but I suppose I could have avoided the ambiguity."[54]

It is at this point we see how far my vision is from that of Richards and why I discount Humboldt. From the viewpoint of science, I see British natural theology as carrying within itself the seeds of its own irrelevance and elimination. If you put God at the beginning as Creator and designer, and from then on keep Him strictly out of the picture, it is all too easy to drop Him entirely. The great Dutch historian of the Scientific Revolution Eduard Jan Dijksterhuis spoke of God becoming a "retired engineer."[55] Once you have a machine in motion, which is the essence of the mechanical worldview, you don't need God anymore. Although machines have purposes, and this speaks

still of organic contrivances like the heart or the eye, the overall purpose of the physical world (if such there be) is no longer a matter
for science. The metaphor of the world-as-a-machine focuses on
its workings and not on its ends.[56] In line with this, Darwin in 1859
was poised to drop God from his world picture, and as he moved to
agnosticism in the 1860s, that is just what he did. God is not needed.
At most we have metaphors, and that, as Darwin kept pointing out,
is just what we have in the rest of science also.

Complementing this, a second reason why I part from Richards is
that—whatever the importance of Humboldt—I still do not see Darwin as a British *Naturphilosoph*. Let me requote what I gave as a paradigmatic example of Darwin-following-Humboldt.

> Among the scenes which are deeply impressed on my mind, none ex-
> ceed in sublimity the primeval forests undefaced by the hand of man;
> whether those of Brazil, where the powers of Life are predominant,
> or those of Tierra del Fuego, where Death and Decay prevail. Both are
> temples filled with the varied productions of the God of Nature:—no
> one can stand in these solitudes unmoved, and not feel that there is
> more in man than the mere breath of his body.[57]

Of course it is Humboldtian. But notice how Darwin is changing it—
twisting it, if you like—in the direction of British natural theology.
It is all about the God of Nature and his productions. This is at least
as much the designer God of Archdeacon Paley as the world spirit of
the Romantics. And this is the point through and through. Again and
again when you get a passage that seems to point to Romanticism,
you can as easily if not more easily point to Paley or a fellow traveler.
Think again of that famous final passage of the *Origin*. It has been
suggested (not by Richards) that it is a reflection of a Humboldtian
passage about nature.[58]

> How vivid is the impression produced by the calm of nature, at noon,
> in these burning climates! The beasts of the forest retire to the thick-
> ets; the birds hide themselves beneath the foliage of the trees, or in
> the crevices of the rocks. Yet, amid this apparent silence, when we

lend an attentive ear to the most feeble sounds transmitted by the air, we hear a dull vibration, a continual murmur, a hum of insects, that fill, if we may use the expression, all the lower strata of the air. Nothing is better fitted to make man feel the extent and power of organic life. Myriads of insects creep upon the soil, and flutter round the plants parched by the ardour of the Sun. A confused noise issues from every bush, from the decayed trunks of trees, from the clefts of the rock, and from the ground undermined by the lizards, millepedes, and cecilias. There are so many voices proclaiming to us, that all nature breathes; and that, under a thousand different forms, life is diffused throughout the cracked and dusty soil, as well as in the bosom of the waters, and in the air that circulates around us.[59]

Well yes, and I don't want to say that Darwin's style and rhetoric owed nothing to this. But my candidate for the influence, Brewster's hymn to the Creator, has a surprisingly large number of words that appear in Darwin's passage. Moreover, whereas Humboldt is talking in a Romantic vein about life infusing through everything—"all nature breathes"—Darwin is talking in a natural theological manner about origins. Just the context and topic of Brewster.

PARADISE LOST

Finally, let me double back and—while admitting the huge influence of Humboldt on the traveling Darwin—point to another, as-yet-unmentioned inspiration for the naturalist of the *Beagle*: the great English poet John Milton. In his *Autobiography*, after telling of the delight in the poetry of Wordsworth and Coleridge (in the years just after the voyage), Darwin wrote: "Formerly Milton's *Paradise Lost* had been my chief favourite, and in my excursions during the voyage of the *Beagle*, when I could take only a single small volume, I always chose Milton."[60] Casually he drops in references to Milton, as in this letter to Henslow: "As for one little toad; I hope it may be new, that it may be Christened 'diabolicus'.—Milton must allude to this very individual, when he talks of 'squat like [a] toad.'"[61] We know also from his reading lists that Darwin went on reading Milton after the voyage. The point simply is that in Milton Darwin found a wonderful creation story that must have fired his imagination.

> over all the face of Earth
> Main ocean flowed, not idle; but, with warm
> Prolifick humour softening all her globe,
> Fermented the great mother to conceive,
> Satiate with genial moisture; when God said,
> Be gathered now ye waters under Heaven
> Into one place, and let dry land appear.
> Immediately the mountains huge appear
> Emergent, and their broad bare backs upheave
> Into the clouds; their tops ascend the sky:
> So high as heaved the tumid hills, so low
> Down sunk a hollow bottom broad and deep,
> Capacious bed of waters.[62]

And then the creation of organisms, superabundant, bursting with life and overflowing the bounds. The world that Darwin discovered in South America.

> And God created the great whales, and each
> Soul living, each that crept, which plenteously
> The waters generated by their kinds;
> And every bird of wing after his kind;
> And saw that it was good, and blessed them, saying.
> Be fruitful, multiply, and in the seas,
> And lakes, and running streams, the waters fill;
>   And let the fowl be multiplied, on the Earth.[63]

And do not forget the climax:

> "Let us make now Man in our image, Man
> In our similitude, and let them rule
> Over the fish and fowl of sea and air,
> Beast of the field, and over all the earth,

And every creeping thing that creeps the ground!"
This said, he formed thee, Adam, thee, O Man,
Dust of the ground, and in thy nostrils breathed
The breath of life; in his own image he
Created thee, in the image of God
Express, and thou becam'st a living Soul.
Male he created thee, but thy consort
Female, for race; then blessed mankind, and said,
"Be fruitful, multiply, and fill the Earth."[64]

Of course, all of this is in a Christian context; but that is my point.[65]

I rest my case. Although he was not exactly offering praise, Karl Marx knew the score. Humboldt and the other Romantics are massively important. They do not do the work that Richards demands of them, nor are they needed to do the work Richards demands of them. We came in with that Pantheon of British heroes, Westminster Abbey, where, next to Isaac Newton, down the aisle from Lyell and Herschel, Charles Darwin lies slumbering. Let us leave him there.[66]

EPILOGUE

Alfred North Whitehead remarked that the European philosophical tradition was but a series of footnotes to Plato.[1] A comparable but more far-reaching observation might be made about Charles Darwin: since the late nineteenth century, intellectual life not only in philosophy but in the sciences and in other areas of cultural significance has been decisively shaped by Darwin's accomplishment. Footnotes to the *Origin of Species* gather like ants to the careless picnic of modern life. Though we, the authors of this book, have major differences in our interpretations of Darwin's theoretical conceptions, we are in little doubt about their impact on the sciences, humanities, and culture more generally. In biology, his ideas have dominated and shaped its history through the last century and a half, and those ideas have recast the understanding of ourselves. In this epilogue, we would like to sketch the major features of that more recent history, noting a few of the fault lines that threaten even this, our small edifice of comity. Of course, substantial changes have occurred in evolutionary theory since Darwin's time, which, we believe, would have surprised—and gratified—him. Nonetheless, even those changes have their roots in his theory, so it would not mischaracterize the biology of the modern period to call it Darwinian, or more precisely, neo-Darwinian. This biology has come to situate human beings into their distinctive place in nature.

HISTORY OF EVOLUTIONARY BIOLOGY
SINCE THE *ORIGIN OF SPECIES*

In 1900, biologists recovered and then, for the next several decades, recast the Moravian monk Gregor Mendel's (1822–84) ideas about hereditary patterns of trait transmission and how to explain those patterns. Particularly at the hands of Hugo de Vries (1848–1935) at the University of Amsterdam and Thomas Hunt Morgan (1866–1945) and his students (Alfred Sturtevant, 1891–1970; Calvin Bridges, 1889–1938; and H. J. Muller, 1890–1967) at Columbia University, the new theory of "genetics" was articulated and developed.[2] Initially it was seen to be a rival to Darwinian approaches to evolution; at the time, the historian Erik Nordenskiöld (1872–1933) even declared Darwinism dead, killed by the real science of the genetics laboratory.[3] In the 1930s, however, biologists began to combine genetic conceptions with Darwinian theory, and a new paradigm gradually emerged, usually called the "synthetic theory" (in America) or the "neo-Darwinian theory" (in England). With the systematic application of mathematics to population structures, the synthetic theory underwent rapid development, especially with the formulation of a principle that specified when a population would be in equilibrium (no change) and how it would change in a predictable fashion under selection pressures. The Hardy-Weinberg principle, named after the British mathematician G. H. Hardy (1877–1947) and the German physician Wilhelm Weinberg (1862–1937), held a position in biology something akin to Newton's first law of motion. With these additions, neo-Darwinism became a theory of the changing genetic structure of populations of organisms. For some biologists, the force of selection has penetrated to the genes; others have held fast to the individual organism as the focus of selection, with genes being carried along by their vehicle, though ultimately responsible for species change through time.

In the early 1930s, the "population geneticists"—notably R. A. Fisher (1890–1962) and J. B. S. Haldane (1892–1964) in England, and Sewall Wright (1889–1988) in the United States—developed the formal theory, and rapidly thereafter a number of naturalists and experimental scientists put empirical flesh on the mathematical skeleton. In England E. B. Ford (1901–88) and his school of "ecological genetics" were important; in the United States, the Russian-born Theodosius Dobzhansky (1900–1975) and his associates—the systematist Ernst Mayr (1904–2005), the paleontologist G. G. Simpson (1902–84), and the botanist G. L. Stebbins (1906–2000)—all contributed to the new theory, if with some crucial differences among them.[4] By the hundredth anniversary of the *Origin* in 1959, the neo-Darwinian synthesis was in place. Or more precisely, rapidly falling into place, for this was also the decade of the double helix, when James Watson (1928–) and Francis Crick (1916–2004) discovered the helical structure of the DNA molecule and therewith provided the foundation for the fine structure of the gene.[5] It seems fair to say, however, that although molecular studies at first appeared threatening to evolutionary theory—a parallel to the beginning of the century when Mendelism posed a danger—before long the insights about the molecular structure of the gene opened new approaches and ways of answering earlier, perplexing questions, especially about the hereditary transmission of traits.

Heraclitus observed that everything flows, and this has been very much true of evolutionary studies. The past century has seen one new finding after another, one new theoretical model after another, one new triumph after another. Let us briefly survey four areas and indicate the kinds of development they have undergone: natural selection theory, paleontology, embryology, and human evolution.

Natural Selection

From the last third of the nineteenth century through the first two decades of the twentieth century, Darwin's chief device of evolutionary change, natural selection, gradually pushed out its principal rival, Lamarckian inheritance of acquired characteristics. Alfred Russel Wallace (1823–1913) in England and August Weismann (1834–1914) in Germany rejected the Lamarckian alternative. Weismann's experiments, on five generations of mice, demonstrated that cutting the tails of mice in each ancestor generation did not cause any shortening of tails in their descendants.[6] These empirical experiments coupled with theories (like Weismann's) that maintained a separation between the germplasm (hereditary material passed from parents to offspring) and the somatoplasm (hereditary material responsible for development of the individual organism) finally extinguished Lamarckism by the mid part of the 1920s (with the exception of a few doctrinaire holdovers in Russia).[7] Some other devices, like genetic drift and structural constraints, complemented natural selection, which became the only major cause the synthetic theory would tolerate. Darwin would have been surprised at the elimination of the inheritance of acquired characteristics from the biologist's repertoire; and he would have been equally astonished at the idea that selection could maintain variation within populations rather than always driving groups to one form or another. The disease sickle-cell anemia is the paradigm case of the retention of variation. Untreated, it leads to an early and painful death. We now realize that, far from selection working to eliminate the gene that leads to this disease—as one might think from an untutored Darwinian perspective—selection is willing to pay the price of a certain proportion (about 4 percent) of disease sufferers who have the gene in the homozygous condition (i.e., they get a double dose of the gene). Evolution pays the price, as it were, because in the heterozygous condition (i.e., when paired with a normal gene), the sickle-cell gene can also confer immunity to malaria, a killer disease in the parts of Africa where the sickle-cell gene was originally found. Selection "balances" the cost of a percentage of very unfit organisms against the benefit of having many more fit organisms.[8]

In the fifty years since the centenary of the *Origin*, increasing attention has been turned to the problem of the levels of selection. In Darwin's view, selection usually wrought its effects on the individual, though in the view of some it can also work at the group level—the reader will have seen that Richards and Ruse fail to agree about this latter issue. In the case of both the individual and the group, selection would have been assumed to work on the whole phenotype (as we would say). But in the 1960s and 1970s, population genetics refocused attention to the gene, arguing that selection principally operated at this most fundamental level. Even complex traits in social organisms came to be understood as resulting from genetic selection. William Hamilton (1936–2000), for instance, formalized a model of "kin selection," demonstrating that an altruistic trait could increase in a population if the trait inclined the bearer to help relatives who also carried the gene for the trait. This was obviously in the spirit of Darwin's family selection. Hamilton, though, was able to take matters to a higher level of sophistication both because of his knowledge of genes and through the application of a shrewd mathematical analysis: roughly, that a trait would be selected if its cost to the carrier were not greater than the benefit to the carrier's relatives in proportion to their degree of relatedness.[9] Or as J. B. S. Haldane more convivially put it when asked if he would jump into a river to save a drowning child: he would gladly risk his life for two brothers or eight cousins (since on average they would represent his own genetic endowment).[10] The American Robert Trivers formulated a complementary model of "reciprocal altruism" and provided various kinds of empirical support for the model.[11] This idea, also grasped in essence by Darwin, has selection promoting quasi-altruistic behavior in organisms because they implicitly expect aid in return. Darwin, though, regarded such expectation as a "low" motive.

Models constructed by population geneticists and social theorists like Hamilton and Trivers, along with the burgeoning amount of empirical evidence that had accumulated by the third quarter of the century, was gathered together in 1975 in one magnificent overview— *Sociobiology: The New Synthesis*—by the American ant-specialist Edward O. Wilson (1929–). The year after, 1976, the English biologist Richard Dawkins (1941–) wrote a famous popular account, *The Selfish Gene*, that summarized a good deal of the individual-selectionist view.[12]

Wilson's work was controversial. It was felt, perhaps with some good reason, that he was insensitive to issues of race and gender. He was also charged with being a rather simpleminded reductionist—something he certainly was not. Criticisms notwithstanding, new studies of the evolution of social structure surged. In the four subsequent decades, a large number of research projects advanced, showing the virtues of an approach to organic nature through natural selection as the principal force shaping animal communities. Societies of dung flies, guppies, lions, naked mole rats, chimpanzees, and many more species were analyzed in this way with incredibly rich results. The principal assumption of this research was that selection operated on individuals or individual-like entities—for example, kin groups. The emphasis on the individual came as a result of the critical analyses, undertaken by George Williams (1926–2010) in the late 1960s, of the then-current models of group selection. Williams discriminated several telling logical and empirical problems in the work of such group selectionists as V. C. Wynn Edwards (1906–97).[13] In the very recent period, however, new models of group selection have been proposed and new experiments conducted—by Michael Wade, David Sloan Wilson, and even by E. O. Wilson—that purport to provide support for the reality of group selection.[14]

Paleontology

Darwin knew that he needed considerable spans of time for species change to occur, but neither he nor his supporters nor his critics knew exactly how much time was needed. Paradoxically, the physical sciences have come to his aid—paradoxical because in the early years after the *Origin* the physicists (ignorant as they were of the warming effects of radioactive decay) were precisely those scientists who contended that there was nothing like the needed time for so slow a process as evolution by natural selection. They drew their evidence from the assumed rate of salt deposition into the sea, heat loss from the earth, and suppositions about the composition of the sun. William Thomson, Lord Kelvin (1824–1907), estimated that the earth would have cooled from its molten state to be able to sustain life only between 98 and 200 million years ago. Though Darwin felt the objection of insufficient time when pronounced by the physicists, he was more wily than one might suspect. In subsequent editions of the *Origin*, he kept ratcheting down the time needed for evolutionary processes, playing on our inability to imagine how long, say, a million years really was and how many generations would turn over in that time.[15]

Now of course we know that the universe is about 13.8 billion years old, that our planet is around 4.5 billion years old, and that life seems to have started almost as soon as the globe got cool enough to bear it, maybe 3.5 billion years ago or even earlier. Today, although significant advances have been made in our understanding of the subject, we are still very far from having a complete story of origins.

Tremendous progress has been made, however, in reconstructing phylogenies that chart the history of life through long periods of time. Expectedly, some areas of the fossil record are better known than others. The evidence of the first 3 billion years of life's history is miniscule compared to the evidence for the last half-billion years. But the story is consistent. We start with primitive forms and work up through the ages to more complex forms. Particularly significant was the Cambrian explosion, more than 500 million years ago, when many of the phyla we find today first appeared in the record. Even more significant is the fact that we do not find anomalies. As Haldane is reputed to have said, there are no rabbits in the pre-Cambrian.

Though Darwin started his professional career as a geologist with a deep interest in paleontology, the *Origin* reveals only a little of his specific phylogenetic concerns—beyond, that is, the descent of fancy pigeons (chapter 1). Of course, in his four volumes on barnacles (1851–54), he traces various species back to their origins in a primitive crustacean rather than in some early mollusk, the class in which Georges Cuvier (1769–1832) placed them. Like his friend Ernst Haeckel (1834–1919), Darwin found the principle of recapitulation to be particularly helpful in establishing the phylogeny of barnacles.[16] In the *Descent of Man*, he did have some speculations about human phylogeny and origins. He presciently supposed that early man developed in Africa and spread to various regions of the world; and he regarded current aboriginal groups as representative of the European's ancestors. Darwin remarked in the *Origin of Species* that the fossil record of phylogenetic development looked like a book with most of its pages torn out, leaving only scattered sentences. Today, many of those pages and sentences have been found and reinserted. The book of life still contains mysteries, but of a more mundane variety.

The past half century has seen a sea change in paleontology with the growth of a whole new subdiscipline of paleobiology, where fossils are treated with all the consideration usually reserved for living organisms.[17] The Hardy-Weinberg law applies as much to trilobites as it does to West Africans. Particularly important is the use by paleontologists (or paleobiologists) of a form of what is known as "reverse engineering," a version of optimization theory. This is as standard a Darwinian way of doing things as it is possible for anything to be. One sits down with a tricky facet of the organic world and tries to puzzle out its meaning from an adaptive perspective, asking how would one design such a structure given a particular problem and the tools and materials to do so. A nice example of the technique occurred when, in 1862, Darwin received an orchid, *Angraecum sesquipedale*, from Madagascar; it had a nectary (the tube at the end of which sweet syrup would pool) of a foot long. He predicted that there had to be a moth on the island with a comparably long proboscis; the creature was discovered in 1903, some forty years after Darwin's prediction.[18]

Darwin used the strategy of reverse engineering repeatedly not only in the particular case of the Madagascar orchid, but as well in the little book on orchids that he penned at the time.[19] The strategy is also fundamental to today's paleontologists, faced as they are with the strange features of brutes from the past. Why does that magnificent dinosaur, the stegosaurus, have weird pointed plates all the way down its back? Various hypotheses have been proposed. Perhaps they arose through sexual selection. Not likely, because both males and females have them. Maybe they were needed for defense or attack. Not likely, because the plates were not anchored to the backbone. Could be that they were used for temperature control, heating the cold-blooded animal in the morning sun and cooling it at midday in the breezes. Much more likely, especially given that the plates seem just like the plates used for heat transfer in air conditioners. There is confirming evidence from the ways in which blood could have been transferred to the plates; although this has led to a rival or supplementary explanation that a large flow of blood could lead to a kind of blushing and hence to a magnificent threat display.

While one function (e.g., heat control) might have been an originating cause, this would not preclude multiple functions for the plates.[20] Asking the adaptive question leads to quite interesting possibilities in the understanding of evolution.

Development

Darwin took development very seriously, especially the relationship between the embryo and the adult organism, a relationship that for him became important evidence of the general theory of descent. In the *Origin*, he utilized the principle of recapitulation (that the embryo goes through the same morphological stages as the phylum had in its evolutionary descent): "Embryology rises greatly in interest, when we thus look at the embryo as a picture, more or less obscured, of the common parent-form of each great class of animals."[21] He became more than a little vexed that critics failed to notice the role of embryology and individual development in his theory. During the period of the synthesis, embryology and development remained of small concern. Most population biologists started with the gene and then jumped straight to the finished organism—from the genotype to the phenotype. If one dips into the popular account given in the *Selfish Gene*, there is much on the relationship between genes (tokens or markers for selection) and the organisms (the frontline troops in the struggle for existence) and not a lot in between. To use a familiar metaphor, organisms were black boxes.

How things have changed! Starting around 1980 or earlier, in the field of evolutionary development ("evo devo" for short), molecular biologists have been tracing the path from the genes, from the DNA molecule, up to the finished organism; and along the way they have come up with some truly staggering discoveries.[22] Perhaps most remarkable is the news that organisms are built along the Lego principle. Put the building blocks together one way and you get the Eiffel Tower; put them another way and you can ride in a Ferris wheel. So in the organic world. Organisms are not built anew from scratch every time. The same molecular components are used, though in different ways, for making a mouse and for making an elephant. Amazing! We might have expected some similarities, let us say between humans and orangutans. But between *Homo* and *Drosophila*?

Obviously all of this is way beyond the expectations of Charles Darwin and his contemporaries. In the 1960s, the human genome was estimated at about 100,000 genes, with simpler organisms assumed to have considerably fewer. Most recently the estimate for the number of human genes has been radically reduced to about 21,000, while the water flea clocks in at 31,000.[23] Obviously sheer numbers of genes do not determine phenotypic complexity; regulation of genes by other genes functions where sheer numbers fail. But are these findings threatening in any sense? Absolutely not! Biological science, like other sciences, continues to advance with better models and better data. The boy who loved chemistry, the young Charles Darwin, would very likely be quite fascinated by these developments in molecular genetics.

Humans

Again, huge advances have been made in the biology of human beings since Darwin. The Neanderthals were known before the *Origin*, though the great German anthropologist Rudolf Virchow (1821–1902) thought they were the remains of brutish Russians who had chased Napoleon back into France.[24] The first proto-human recognized as such was discovered in the early 1890s, Java Man, found by Haeckel's protégé Eugene Dubois (1858–1940).[25] Then, in the twentieth century starting with the Taung baby (an australopithecine or southern ape), discovered in South Africa by Raymond Dart (1893–1988), the fossils gushed forth. Great credit goes to the indefatigable Leakey family working in East Africa, although the find of the century was surely "Lucy," a little hominin just over three feet tall, with a chimpanzee-sized brain (about 400 cubic centimeters as opposed to our 1,250 cubic centimeters), walking on her hind legs, and more than 3 million years old. Molecular biologists who showed that estimates of a 10-million-year separation of apes from the human line were initially far too conservative. We split about 5 million years ago and, whatever the appearances, it seems that we humans are more closely related to chimpanzees than the chimpanzees are to the gorillas.[26] Molecules again have recently been in the news as researchers have been able to collect ancient DNA and have answered that intriguing question: Did Neanderthals and *Homo sapiens* engage in any trysts? The answer apparently is yes, since we carry some few Neanderthal genes, the results of occasional interbreeding till the Neanderthal line ran out, about 40,000 years ago.[27] Old-fashioned fossil hunting has turned up one of the most remarkable of all discoveries, the little hominin, *Homo floresiensis*, a creature dubbed "the hobbit," from Indonesia. *H. floresiensis* seems to have lived as recently as 12,000 years ago, and perhaps descended as an isolated group from *Homo erectus*.[28]

All of this would be new to Darwin, although nothing of great conceptual surprise or worry. Darwin, like all of his contemporaries, assumed that there was a progressive path to humans, an assumption now challenged. But the rise of human brainpower, if unexpected, is not totally inexplicable, as our ancestors used their new adaptation to hunt and forage and probably to fight fellow species members. What about those brains? It is clear that Darwin had little time for a blank-slate theory of intelligence. He thought that the ways in which we reason and act, particularly act morally, were (to use a modern metaphor) part of the brain's software. This is the beginning assumption of a whole class of modern-day Darwinians, the "evolutionary psychologists." They argue that our reasoning and our moral sense are both part of our genetic heritage. They would agree with Darwin that such things as basic reasoning—logic, mathematics, and above all language abilities—are part of our innate legacy as given to us by evolution through natural selection. Where they would go beyond Darwin, perhaps, is in arguing that some of the quirks of our reasoning are likewise explained by selection, though against a much earlier environment, one rather different from ours.[29]

Theories of moral evolution have been the subject of much scrutiny in recent years. Experimental studies of apes and monkeys, for instance, have shown that these creatures engage in what appears to be protomoral behavior (e.g., consolation for injuries suffered) or even apply moral norms (e.g., negative expressions when elemental standards of justice are violated).[30] Survey studies of human beings—across nationalities, culture, religion, and education—indicate common, deep-seated intuitions about morally appropriate behavior. Paradigmatic are the trolley problems. When subjects are presented with these problems their resolutions indicate a general consensus about circumstances under which one feels a moral obligation to save five people at the expense of one and about other circumstances in which such a trade-off feels morally wrong.[31] Darwin, of course, pioneered the proposal that selection has instilled in human beings deep-seated instincts about morally appropriate behavior.

HUMAN CONSCIOUSNESS

Theodosius Dobzhansky remarked that "nothing in biology makes sense except in the light of evolution," and the evolution he had in mind was Darwinian.[32] Beyond the confines of empirical biology, however, does Darwinism have implications for the traditionally metaphysical questions concerning human consciousness? We believe it does, though such implications cannot be completely decisive in answering these questions.

As every moderately literate individual knows, a large number of positions have been taken on the relationship between human mind and human brain. We don't propose to discuss the multitude of these positions, which multiply like inflating universes. (C. D. Broad in 1925 enumerated some seventeen different mind-body theories; and others have emerged since.)[33] We will, rather, consider the most prominent ones, with a word or two about their varieties. They fall under four somewhat overlapping categories: dualism, materialism, monism, and emergentism. In what follows, we will assess these positions in relation to evolutionary theory.

Cartesian dualism contends that mind and brain are separate entities. Descartes (1596–1650) believed that the properties of each were radically different, and so had to be located in separate kinds of substance: a mental, thinking entity and a material, extended entity. Descartes's view had its roots in Plato and subsequent medieval philosophers. Many religious conceptions of human beings suppose some kind of dualism, since the doctrine holds the promise of life after death. The great Renaissance anatomist Vesalius (1514–64) expressed the common view, which he inscribed in his classic depiction of the human body: "Vivitur ingenio, caetera mortis erunt" ("Mind lives, all else is mortal").[34] Substance dualism is an all or nothing kind of thing. Even Erik Wasmann (1859–1931), a Jesuit evolutionist of the late ninetcenth and early twentieth centuries, whose views became canonical among Catholic theologians and philosophers, required that at some moment during the evolution of man's body, God infused the soul, which carried the human mind.[35] Darwinism precludes the endorsement of such miracles. Mind, in the Darwinian view, is a historical accomplishment, gradually becoming more complex in the hominid line, but with mental and psychological traits found at various levels in the collateral branches within the animal kingdom.

Materialism comes in several varieties. The most radical position is that of Daniel Dennett, who argues that consciousness, human mind, simply does not exist. All that exists is the human brain, which consists ultimately of particulate matter. By this strategy, Dennett tries to avoid what has become known as the "hard problem" in the philosophy of mind, namely, the relationship between brain and a unified consciousness, a self. He virtually reverses the Vesalian epigram. In order to make his argument stick, he has to contend that qualia—the red and green, the rough and smooth, the loud and soft, the hot and the cold of our experience (those purported objects of consciousness)—do not really exist.[36] Since not all of our brain activity produces apparent phenomenal awareness, Dennett must suppose there is no real distinction between brain activity that produces apparent consciousness and that which does not. After all, in his view, unified consciousness does not exit. But virtually all of neuroscience makes this fundamental distinction between conscious experience and brain activity; indeed, fMRI investigations of the neural substrate of consciousness suppose the existence of consciousness to guide and verify features of neural activity. Dennett's arguments are as ingenious as they are unconvincing.

Another version of materialism maintains that mind is simply the behavior or function of the brain, the brain's activity. The distinction between brain states that function without consciousness and those that produce consciousness must lie in different kinds of activity. But either mind activity is like ordinary neural activity that doesn't have conscious accompaniment—then it's essentially a version of the radical type of materialism—or that particular activity of the brain is different. Yet the only difference could be that it produces phenomenal awareness. In regard to the first alternative, one could imagine a race of apparent human beings whose brains work as ours do but without phenomenal experience—the zombies that populate contemporary films and TV. But such creatures would be distinctively different from us—the difference being consciousness. So materialism must recognize a particular kind of neural activity that gives rise to consciousness; and that activity is identified precisely by its production of consciousness. But then consciousness must be something more than the ordinary activity of the brain. This leads to the position of epiphenomenalism.

So human brains (and brains of higher animals) do produce phenomenal conscious states with their blazing array of qualia, and these cannot be simply identical with ordinary brain states. The epiphenomenalist holds that states of consciousness produced by human brains are simply inert; they are like, say, the redness of blood—a property that has no function but is only, as it were, carried along by the hemoglobin molecule, which does have a function. So this version of materialism maintains that consciousness is a property produced by the human brain but one that has no causal function. Thomas Henry Huxley endorsed this kind of mind-brain theory. Huxley argued that brain activity had two different kinds of effect: it caused physiological and behavioral actions, and it also caused consciousness states. But the conscious states, in their turn, were inefficacious; they did no work.[37] William James (1842–1910) produced a telling Darwinian argument against this version of "steam-whistle" epiphenomenalism—so called because consciousness is supposed to have no more effect on actions than the whistle on a locomotive— and he was followed by the philosopher Karl Popper (1902–94) and the Nobel laureate and neurophysiologist John Eccles (1903–97), both of whom subscribed to James's argument.[38] If consciousness slowly evolved from its first glimmers in early animals through proto-humans to modern man, like all properties it must have evolved under the aegis of natural selection. But then if selected, it must have a use; otherwise selection would have eliminated such a costly but unproductive trait, leaving only zombies in both the animal and human lines. If consciousness has a use, then it must do work beyond what the naked brain performs.[39]

The counterargument of a convinced epiphenomenalist might be that consciousness is simply a necessary concomitant of brains (like the red of the hemoglobin molecule). But the recalcitrant epiphenomenalist must recognize that consciousness, along with brains, has become ever more complex over evolutionary time, that the features of consciousness show developmental integration (of perceptions, memory, deliberation, feelings, etc.). Yet there is no reason to suppose that the integration of features of consciousness should match the integration of brain parts. If it were simply the case that a set of brain parts produced a concomitant set of conscious parts in the way hemoglobin produces the color red, there would be no reason to expect the conscious system to be anything but a hodge-podge, a Jackson Pollock of the mind. Brain may secrete thoughts as the liver secretes bile, as Darwin supposed, but the bile is real and does work, and analogously, so do thoughts.

At the end of the eighteenth century and through the early part of the twentieth, Spinoza's neutral monism gained a second wind. Goethe, Haeckel, Wallace, Spencer, Schelling, Mach, and Russell all endorsed a neutral monism of varying differences. For Baruch Spinoza (1632–77), mind and matter were properties of a single underlying substance; their coordination was assumed because of their deep connections. Herbert Spencer (1820–1903) supposed that each bit of matter came with a bit of mind as its concomitant; so there were as many substances and mini-minds as there were atoms. James and Bertrand Russell (1872–1970) contended that experience (or sensation) provided the neutral stuff, out of which we construct mind or matter: the red of the ribbon in our experience could be construed now as a feature of the physical world, now as a perception of the mind.[40] Many problems attend the various types of neutral monism that we've briefly mentioned, but two in particular stand out. In the case of Spencer and Wallace's version, William James objected that this was essentially a mind-dust theory: the little bits of consciousness, even if hovering around a brain, would have no central unity, no self to integrate and coordinate them.[41] This was precisely James's own problem, and Russell's as well. James's version of neutral monism held: "That entity [consciousness] is fictitious, while thoughts in the concrete are fully real. But thoughts in the concrete are made of the same stuff as things are."[42] Yet, what is the nature of that entity or mind doing the constructing of neutral experience, constructing this red as a mental perception at one time, a physical property at another? Moreover, one could ask whether objects exist in the universe that have never been experienced or sensed. Most sciences, especially evolutionary biology and physics, assume the existence of kinds of objects that have passed through no one's stream of thought. What justifies the assumption, perfectly ubiquitous, that such objects nonetheless exist and wait to be discovered?

Emergentism is the last view we would like to consider, and the one most comfortable with Darwinian theory. Emergentism holds that consciousness has gradually arisen as the nervous system has become ever more complex during the evolutionary history of organisms. This means that while humans possess the highest form of consciousness and rational capacity, other animals are not completely bereft of mind. This doesn't mean that if evolutionary theory in general is true, that mind must be its gradually evolved product, but it certainly makes emergentism more likely. Neo-Darwinians claim to be able to explain salient biological traits, and if mind is a biological trait—an emergent property of brain—then it should be explicable in terms of neo-Darwinian theory. If the rise of mind cannot be explained by evolutionary biology, then that might well cause us to hesitate about endorsing Darwinian theory. Thomas Nagel, a distinguished philosopher of mind, has recently argued that mind cannot be explained by evolutionary theory and thus evolutionary theory, as it stands, is radically incomplete. Something is missing. But that is to say, in Nagel's tendentious interpretation, evolutionary theory "is almost certainly false."[43]

Nagel has based his objections fundamentally on what he takes as the inability of evolutionary theory to solve the hard problem in the philosophy of mind, namely, explaining the features of consciousness using only the resources of evolutionary naturalism. We will enumerate several of the arguments he makes along the path to his most basic objection. His first salient argument is that an evolutionary construction of mind makes human reason unreliable: "Evolutionary naturalism provides an account of our capacities that undermines their reliability and in doing so undermines itself."[44] If our mental faculties have evolved to deal with very practical problems of life, they "do not warrant our confidence in the construction of theoretical [i.e., scientific] accounts of the world as a whole." This objection, of course, supposes distinctly different kinds of human mental capacities: those directed toward the everyday beliefs about the world and those directed to the establishment of scientific beliefs. But there is no justification for this distinction. The whole history of science testifies to the gradual application of common reasoning ability, exercised in the business of life, in the construction of scientific conceptions. Part of the repertoire of our cognitive dealings with the immediate world are various mental instruments for correcting initially faulty impressions: we certify that the stick in the water, though it looks bent, is really straight by feeling it, pulling it out of the water, noticing that all such objects in water look bent from the same angle. Those of our ancestors who lacked these corrective mental instruments, these checks on immediate perception, fared poorly in the struggle for existence. Is the correction of more abstract, scientific beliefs, fundamentally different? Our cognitive efforts in science, as the history of science makes clear, have been unreliable until yesterday; the science of the previous generation has been superseded, as we presume this one will be as well.

Getting closer to the heart of the matter, Nagel objects to "Naturalism," the program that regards things in the world as "physical things" and subject to natural law.[45] He believes the naturalism that is endemic to modern evolutionary biology fails to explain consciousness, because consciousness is not a physical thing. Contemporary physics, however, is willing to consider the ultimate foundation of reality in vibrating mathematical strings and to postulate the existence of "dark matter" and "dark energy," which seem to have properties radically different from what used to be thought of as the standard constituents of matter. Many philosophers who try to seal hermetically the realm of consciousness from the realm of matter still harbor Newtonian notions about matter, regarding it as constituted by those hard, glassy impenetrable particles that do seem so very different from the elements of consciousness. But contemporary naturalism, given the scope of modern science, seems ready to encompass any phenomena that can be comprehended and studied by human reason. The laws (deterministic or statistical) that now exist in the sciences presumably include the laws of population genetics, laws of learning in psychology, and sociological generalizations about small group phenomena. Few would deny that in psychology, reliable laws of conscious perception have been discovered—for example, color constancy—or that laws of cognitive development have been instrumental in understanding the behavior, say, of young children. Aren't these examples of an ingression of naturalism already into the domain of consciousness? The future development of the biological understanding of conscious phenomena appears wide open. What would be precluded under this more contemporary conception of naturalism would be supernatural entities whose behavior would in principle be inaccessible to human reason. Just as physics and biology endorse the emergence of new properties in the universe over evolutionary time spans, so there should be no a priori exclusion of the emergence of mind, as a new property of the universe.[46]

Another strategic way Nagel attempts to undermine evolutionary theory is by maintaining that its explanations are radically incomplete:

> Selection for physical reproductive fitness may have resulted in the appearance of organisms that are in fact conscious, and that have the observable variety of different specific kinds of consciousness, but

there is no physical explanation of why this is so—nor any other kind of explanation that we know of.[47]

Nagel offers several variations on this theme of the insufficiency of evolutionary accounts: if mind has evolved through natural selection, how does this occur? More generally, exactly why does this kind of brain bring with it this kind of conscious mind?[48] Of course, we have some clues to the answers to these questions. Dogs and their evolutionary progenitors found advantage in detecting prey, which often would be hidden in high savanna grass, by an acute olfactory sense. So evolutionary theory does give an account of why we, who are principally visual creatures, have a less keen sense of smell compared to our dogs; and this is physically revealed in our proportionately smaller olfactory area of the brain. Evolutionary theory does, therefore, give an account of the complexity of certain brain areas associated with complexity of phenomenal experience: both resulted from selection pressures.

Nervous tissue, of the sort that constitute brains, arose through natural selection in the deep past; and it would appear that the first glimmers of consciousness have as well. But Nagel wants more: evolutionary theory "would have to offer some account of why the appearance of conscious organisms, and not merely of behaviorally complex organisms, was likely."[49] Admittedly, we are uncertain how the first glimmers of consciousness arose as an emergent phenomenon from nervous tissue. But then, we also don't quite understand how nerve cells evolved from other kinds of cells; yet that doesn't mean that natural selection as an explanation is deficient. By parity of reasoning the same holds for the gradual emergence of consciousness: a less than full explanation does not mean that evolutionary theory is false or mistaken—it may suggest, rather, that cell biology has not yet found its resting place.

Nagel is willing to push the demand for further explanation to the brink: How can one explain, he asks, the transition from non-life to life? He, following the intelligent designers, argues in Zeno-like fashion. If one can demonstrate, say, a transition between major animal groups, for example, between reptiles and mammals based on paleontological evidence, one can still ask: But wait, what about the intermediate stages? How were they made? And if a tentative answer is proffered, then one can pursue the ever smaller gap: What about the in-between step? This quest will never be satisfied. Why is grass green? Well, because of the light reflected off a surface that is absorbed by the retina at length of about 450 nanometers. But why is that length absorbed? Well, because of the kind of protein in the cones. But why does the protein absorb light in that range, and so on? At some point one must simply say that A causes B and no further explanation is possible, at least for the time being. But this is the situation with all of science, and so it is unreasonable to ask of evolutionary theory what is not asked of other sciences.

Natural selection theory, despite Nagel's appeal to Zeno's logic, does explain why consciousness exists: because it more effectively responds to environmental problems than automatisms can. Creatures low on the scale of consciousness must compensate for insufficient consciousness by hyper-reproduction, which is why the world is bulging with beetles; but as consciousness increases, organisms can be more flexible in their responses to the environment, and so reproduction declines.

The last redoubt of objections to which Nagel repairs is that of normative judgment both in our cognitive commerce with the world and with our moral commerce with other people. In the main body of this book, we have taken on the problem of human moral behavior, arriving, however, at different analyses of the role of evolution in determining moral capacities. But we are in agreement about the nature of consciousness and the Darwinian reply to Nagel's objections.

Nagel is willing to concede that sensation and perception of objects might yield to a natural selection account. But he thinks reason operates in a completely different way. In "ordinary perception, we are like mechanisms governed by a (roughly) truth-preserving algorithm." But in the case of reasoning, "something has happened that has gotten our minds into immediate contact with the rational order of the world order."[50] At this point, Nagel begins to light the candles, looking to peer beyond the natural world to discover in us an ability to come face to face with a different kind of realm than the natural. Need we go in that direction? Isn't there a more mundane explanation available? The algorithms of reasoning, patterns of cognitive activity that enshrine such imperatives as avoiding inconsistency, of recognizing the principle of "dictum de omni, dictum de nullo," and of proceeding systematically seem necessary requirements of social creatures that have begun to use language. And like the capacity for language, such principles would have been selected for over long periods of time. For a Darwinian, these principles are not the result of direct insight into the deep structure of reality, as Nagel seems to believe humans capable, but the result of pragmatically dealing with the world. The history of science does not indicate that researchers have utilized a preternatural insight into truth; rather that they have stumbled toward the refinement of cognitive instruments and have used them to comprehend a puzzling world.

What is missing from the Darwinian scenario, Nagel argues, is an appreciation of the teleological character of the evolution of life, of consciousness, and of reasoning ability. Intelligent designers also maintain that these processes of life, consciousness, and reasoning are teleologically structured; but they have a way of anchoring the notion of teleology: divine intention determines the effective aim and cause of the process. Nagel, being an atheist, rejects the intervention of a divine mind. But without an all-powerful intentionality fixing the end goals of life, of consciousness, and of reasoning capacity, what does it mean to suggest these processes are goal directed, teleologically structured?

A teleological causal analysis implies that a developmental process—for example, the gradual evolution of reasoning ability—has its further structure determined by the goal of the process—for example, human, scientific rationality—and that each of its stages is oriented toward that goal. If I decide to construct a model airplane, the plan I have in mind guides each of the steps that I take: the choice of the wood, the shape of the fuselage, the length of the wings, and so on. Without a determining mind behind organic processes, operating according to an idea, a plan, it's difficult to understand what kind of teleology Nagel is proposing. But there is one kind of teleology that he seems to dismiss—Darwinian teleology—which is perfectly adequate to the requirements he specifies.[51]

The general end of organisms, that which constitutes their goal, is reproductive success: that is the ultimate criterion or standard governing evolutionary processes. The environment and the previous state of the organism determine how that goal might be realized; they provide the structure of development leading to the goal. Consciousness and rationality might be used for many ends: solving mathematical problems, tracking a comet, or taking delight in a sunset. But these fundamental human capacities must have ultimately arisen for facilitating reproductive success. Darwinian natural selection was, after all, the full account that Nagel was searching for.

RELIGION AND GOD

In his *Autobiography*, Darwin testified that at the time he finished the composition of the *Origin of Species*, he still believed that something like mind governed the universe.[52] The *Origin* itself demonstrates (at least for one of us) that Darwin's theistic belief helped structure his views about natural laws as secondary causes, with the divine mind as the primary cause. That belief gradually dissolved in succeeding years into an uncomfortable agnosticism, with Darwin declaring a few years before his death: "In my most extreme fluctuations I have never been an atheist in the sense of denying the existence of a God.—I think that generally (and more and more so as I grow older), but not always,—that an agnostic would be the most correct description of my state of mind."[53] In the *Descent of Man*, Darwin did, slyly, indicate what he thought to be the source of conventional religious attitudes. He reckoned that the aboriginal's belief in spirits and gods was comparable to the state of his little dog, who chased after a wind-blown parasol, assuming some imperceptible creature was moving it along. The "primitive" also attributed strange natural occurrences to the actions of unseen agents, and, like his dog, displayed a religious reverence for the invisible master. Darwin hastened to add, he was not touching on the question whether an omnipotent God actually exists, which "has been answered in the affirmative by the highest intellects that have ever lived."[54] Of course, Darwin was not only touching on the question, he was shaking its foundations.

Subsequent biologists, E. O. Wilson, for example, have argued that belief in God and the rituals of religion arose not only for reasons of the kind supposed by Darwin, but also for reasons of group solidarity. Religion might function to encourage both altruism and subordination to a group leader. There might even be a kind of cultural competition among religions, with one group's religion proving successful insofar as it promoted dominance over other groups.[55] Such functional evolutionary arguments, of course, are corrosive of traditional religion and dogmatic belief. Such acidic tendencies can be detected at work among the tribe of biologists.

In 1914, the psychologist James Leuba discovered that among scientists surveyed about 42 percent expressed a belief in a personal God (i.e., a being who would answer prayers); among elite scientists (those mentioned in *American Men of Science*), that figure fell to 35 percent; and for elite biologists, 17 percent.[56] In 1996, Larson and Witham repeated Leuba's survey, using more modern techniques of polling, and found that the level of belief in a personal God among scientists at large held pretty steady, about 39 percent. However, among elite scientists (members of the National Academy of Sciences) that figure fell to 7 percent. Among elite biologists, only 5.5 percent professed belief in a personal God.[57] It would appear that biologists of considerable acumen, like Darwin of later years, have regarded their science incompatible with a belief in a personal God.

Is there yet a way of squaring belief in a supernatural power with evolutionary theory? Certainly not with the beliefs of so-called scientific creationists. They have simply denied the great age of the earth and rejected the evolutionary transitions for which fossils and well-established dating techniques provide strong and comprehensive documentation. The best evidence shows, for example, that the cynodonts, a clade of therapsid reptiles that appeared about 275 million years ago, gave rise to the mammals about 50 million years later.[58] Our own line shows the same gradual transitions: from *Australopithecus afarensis* ("Lucy"), at about 3.5 million years ago, to *Homo habilis*, which had a brain size of about half ours and lived 2.5 million years ago, through the various lines of *Homo erectus*, and side lines of *Homo neanderthalensis*, to *Homo sapiens sapiens*, our species, which appeared about 43,000 years ago.[59] Not to forget, of course, the curious diminutive *Homo floresiensis*, likely a surviving branch of *Homo erectus* in Indonesia, which seems to have lived well into the time of modern man.[60] Quite clearly, these well-confirmed conclusions from recent evolutionary biology run counter to the beliefs of fundamentalists-creationists, who account for four in ten Americans. Thus a significant part of the American public are beyond the circle of science—and reason—at least on this topic.

In a Gallup poll of 2014, 42 percent of U.S. adults thought that species were created by God much as we now see them; 31 percent agreed that humans have evolved, but guided in the process by God; and 19 percent held that the neo-Darwinian process alone was sufficient for the evolution of human beings. The results might be contrasted with a similar poll in 1982, which yielded 44 percent, 38 percent, and 9 percent, respectively. In some thirty years, the percentage of Americans holding to the stark biblical story has changed very little, while the neo-Darwinian account seems to have moved those who had accepted theistic evolution into the category of evolution neat. It's hard to say what has caused this minor shift: perhaps more Americans receiving higher education, perhaps the failure of Creationism in the courts, perhaps the publicity given such prominent atheists as Richard Dawkins, Daniel Dennett, Samuel Harris, and Jerry Coyne—biologists and philosophers of biology all. Does this shift suggest the power of the evolutionary cum atheistic argument? And should we conclude that neo-Darwinian theory is strictly incompatible with religious belief?

The shelves of bookstores and the nudging suggestions of online providers are now replete with books arguing the stark opposition of Darwinian science to any religious belief. Daniel Dennett uncorked the current stream with *Darwin's Dangerous Idea* (1995). That stream, now at gentle flood, includes such books as Sam Harris's *The End of Faith: Religion, Terror, and the Future of Reason* (2005), Richard Dawkins's *The God Delusion* (2008), and Julien Musolino's *The Soul Fallacy: What Science Shows We Gain from Letting Go of Our*

Soul Beliefs (2015). Some authors, like Dennett, have again returned to the issue—in his *Breaking the Spell: Religion as a Natural Phenomenon* (2007)—and some, like Victor Stengler, seem obsessed: *God, the Failed Hypothesis: How Science Shows That God Does Not Exist* (2008); *God and the Folly of Faith: The Incompatibility of Science and Religion* (2012); *God and the Multiverse: Humanity's Expanding View of the Cosmos* (2014), and so on. The arguments in these books have a vintage ring to them. Most of these arguments can be found in simpler form in the pellucid essay by Bertrand Russell, "Why I Am Not a Christian" (1927), or earlier and with more colorful aplomb in Voltaire's *Dieu et les hommes* (1769). It's worthwhile examining the tenor of these arguments as they appear in the very recent book by the distinguished evolutionary biologist Jerry Coyne, in his *Faith vs. Fact: Why Science and Religion Are Incompatible* (2015).

Coyne directs his passionate attack against one particular strand in the science versus religion debate: arguments that try to preclude conflict by relegating religion strictly to the emotional and moral, while keeping science the preserve of the rational and empirical. Hence, in this lamented scenario, emotional states or moral claims about what ought to exist can lie down with any empirical claims about what does in fact exist. The most notable effort to launch this kind of accommodation is that of the late Stephen Jay Gould. Gould did not distill this division from the actual practice and pronouncements of theologians and scientists through the ages, but simply declared what religion and science should be. From such a subjunctive declaration, conflict is, of course, avoided:

> Science tries to document the factual character of the natural world to develop theories that coordinate and explain these facts. Religion, on the other hand, operates in the equally important, but utterly different realm of human purposes, meanings, and values—subjects that the factual domain of science might illuminate, but can never resolve.[61]

As Coyne is quick to point out, this stipulative definition of religion is violated by most of the earth's religions. Most religions make existential claims, at least by asserting God's existence; beyond that, many religions have proposed miraculous interventions in the world. Coyne doesn't mention it, but from the science side, values flow across any proposed boundary; that is, science itself is grounded in values. Thus, normative standards for acceptable knowledge-claims and for appropriate methods have formed the framework of science through the ages. Darwin himself often transgressed Gould's claim of separate and nonoverlapping magisteria when he invoked God as a primary cause in the *Origin* and later in the *Descent* undertook an evolutionary explanation of moral values. Darwin's efforts would have been arbitrarily precluded by Gould's irenic resolution to the science-religion struggle. Gould did not discover two nonoverlapping magisteria; he merely promulgated them as a normative resolution to the obvious conflict of science and religion. He made accommodation by fiat.

In dealing with some of the factual claims of prominent religions, such as the reality of Adam and Eve (whose existence is implicated in doctrines of original sin and the coming of Christ), Coyne handily arrays the findings of recent genetics. He also easily explodes the "genetical theories" of the Angel Moroni, who, according to Mormon doctrine, inscribed in the golden tablets that Native Americans originated in the Middle East. Insofar as these assertions of religious doctrine are taken straightforwardly as factual, science stands against them with compelling evidence. Against theistic evolution (of the kind Darwin seems to have entertained while writing the *Origin*), the countervailing considerations require more finesse.

Coyne himself at times slips into argument by fiat: "theistic evolution has been completely rejected by scientists."[62] What he must mean is that theistic evolution (i.e., evolution directed by divine power) has been rejected by a fair number of scientists, certainly not all. In his book, he cites several theistic evolutionists who have exemplary credentials as scientists.[63] A principal charge he brings against theistic evolution is the incompatibility with what we know of genetic change, particularly its contingency, which would argue against the theistic assumption of the inevitability of man. While this evidence does not tell against the sheer logical possibility of a divine power behind the scenes orchestrating the whole, it makes such assumptions look like an irrational stretch, an instance of faith overcoming likelihood. To make theistic evolution work, independent evidence for the existence of such a divine power would be required, some rational grounds that might provide a footing for faith. Coyne considers the kind of view advanced by Conway Morris and Kenneth Miller, both first-rate scientists and Christians.[64] These scientists argue that the phenomenon of convergent evolution suggests the existence in the deep past of an adaptive space favorable to greater intelligence, a space of the sort that would inevitably lead to human beings. But without other examples of convergent evolution of intelligence, the argument has no power: that humans have evolved logically implies only that the conditions of possibility were antecedent; thus the argument from adaptive space seems only to mean that since humans did evolve, they could have evolved. That hardly appears to supply the rational footing wanted.

Was human evolution contingent, as Coyne suggests, or fixed in advance, as Morris and Miller seem to believe? Coyne recognizes that to call evolutionary processes "contingent" does not mean, at a first level of analysis, "undetermined" but only "unpredictable." Yet Coyne proposes that at a deeper level the processes of evolution do lie on radically contingent grounds; he suggests that if quantum indeterminacy ultimately yielded these evolutionary processes, then evolution would be fundamentally contingent. When, however, Miller used quantum mechanics to argue that God might be lurking in the indeterminacy of subatomic processes, Coyne dismissed the supposition with the remark that Miller was "camping on the outskirts of creationism."[65] Both sides quote quantum mechanics for their own purposes.

Here it would be helpful to bring in David Hume, one of Darwin's favorite philosophers. Hume contended that "a wise man . . . proportions his belief to the evidence."[66] And since the sciences have proved to be a coherent, interdependent framework for understanding, together they stand as powerful evidence against any assumptions of miraculous interventions into the natural world, interventions that might covertly and undetectably slip into the natural world, changing this, modifying that, or orchestrating everything from an external vantage. To allow this would be to give up rational control of an enterprise that has proved extremely successful, and has been so only through control from within. Control from within occurs as rational actors continue to make discoveries warranted by evidence and to adjust the findings of the sciences to one another—it is an engine propelled by its own energies. So, following Hume's dictum, it would not be the part of wisdom to allow a remote possibility to outweigh a weighted probability. To allow miracles would be to thwart the rational character of science and destroy its integrity. At most, one can say that theistic evolution is a logical possibility, but that the evidence is heaped against it.

The new "natural theologians," as Coyne describes them, have attempted to caution against the liabilities of scientism, the belief that science is the exclusive mode of knowing. Coyne thinks such cautions are based on a faulty assumption, because science is in fact the only way of knowing the world. He challenges professors of literature and literary critics "to give me examples of truths actually *revealed for the first time* by literature."[67] But is this challenge so hard to meet? Keats declared that he had only heard rumors about the fabled realms of gold but never understood them till he "heard Chapman sing out loud and bold." Did he and others not then acquire some new knowledge and for the first time? Many an ancient Greek adolescent, as well as modern college student, has perhaps understood fragments of war, the elements of savage revenge, of blind rage, of the prideful stupidity of generals, of a father's tender love, and of a woman's passion for a beautiful but unworthy man—but did any see the thing whole before the Homeric songs orchestrated the intricate relationship of these elements? Those songs provide real experience of psychological and social behavior that generations of readers have found true, and have modeled their own behavior on. In a comparable way, there is scarcely a feature of Darwin's theory that had not been fragmentally known before he wrote, though something new was revealed in the "long argument" of the *Origin* that wove those features together.

Reciprocally, science makes use of metaphor—the special provenance of literature. When Darwin exclaims that "we behold the face of nature bright with gladness," he calls the reader's attention, by contrast, to the great life-destroying struggle that occurs behind a happy mask and cautions against being deceived by surface appearances.[68] "Natural selection" is itself a metaphor that structures the entire theory in unspoken ways; our different analyses in the body of this book indicate some of those possible ways. Models in science are in fact metaphors, instruments for understanding a phenomenon by employing more tractable considerations to get at the less tractable. The Cambridge philosopher Mary Hesse distinguished three types of analogies packed into the metaphors of science: a positive analogy, by which the model and the natural phenomenon are assumed to be alike; a negative analogy, by which they differ; and a neutral analogy, which suggests areas of investigation wherein one might find either further similarities or differences.[69] The positive analogy is pedagogically useful; the neutral analogy is useful as a guide to further research. Such metaphors in literature and science operate to acquaint us with aspects of the world hitherto unnoticed. Without the literary device of metaphor, Darwin would not have been able to stake his scientific claim, to make it plausible to his readers and to himself.

Bertrand Russell distinguished two kinds of knowing: knowledge by description, which depends on inferences drawn ultimately from foundational knowledge, and knowledge by acquaintance, which is that foundational knowledge.[70] The latter is immediate, noninferential awareness, an observation of some feature of the world, as when Galen, for instance, first dissected an eye and noted the five layers of the cornea. Without knowledge by acquaintance there can be no theoretical, inferential knowledge. But is knowledge by acquaintance in the sciences fundamentally different from the kind of knowledge by acquaintance illuminated by an arresting literary metaphor or graphic line of poetry, the kind that gives us for the first time immediate knowledge? Hasn't the last line of Randall Jarrell's "Death of the Ball Turret Gunner" jolted any number of naive high school students out of an obsession with the faux glory of the warrior? The sciences are deposits of different ways of knowing, some of which intersect with the kind of aesthetic awareness that literature provides. Salutary, then, is the caution that science is not the narrow and only road that leads to revelations about the world. To state this, of course, would not deny that physics, biology, and the social sciences have discovered quite reliable ways of knowing, just not the only or exclusive ways.

But does religion provide knowledge of any kind compatible with science? Of course, that depends on what you mean by religion. Coyne tells of many stupid and egregious decisions justified by religion—the Christian Science parents, for example, who allowed their daughter to die of bone cancer, choosing prayer over medical treatment.[71] No rational and moral person would defend the parents in their delusion, but likewise no rational and moral scientist should have sanctioned the Tuskegee experiments. Neither science nor religion makes bad decisions, individuals do. Scientific knowledge— certainly including evolutionary theory—gives actors the instruments to make informed, rational, and moral decisions. Does religion of any sort perform a similar function? Consider the view of Friedrich Daniel Schleiermacher (1768–1834).

In 1799, Schleiermacher, a member of a loose confederation of Romantic poets, philosophers, and scientists, published a tract entitled *On Religion: Talks to the Cultured People amongst Its Despisers*.[72] The despisers were very much like Coyne and friends, completely dismissive of religion. Without detailing the deep features of what Schleiermacher takes to be the kind of intuition—knowledge by acquaintance—grounding religious response, suffice it to say it is a feeling of dependency in light of the infinite character of the universe, a universe that operates according to scientific principles but yet lies still beyond the grasp of such principles, a mysterious beyond that seems to have no end. J. B. S. Haldane, conceptually arrested by this aspect of science, suspected that "the universe is not only queerer than we suppose, but queerer than we can suppose."[73] But we do, in our way, come to grips with nature. In the history of science, mysteries have gradually given way to human reason, but continue to reveal an ever greater vastness of the unknown—beyond our now conventional understanding of matter lies, perhaps, its further ground in mathematical entities called strings and beyond that in an imponderable dark matter and energy, all of which cannot but yield a feeling of dependency and respect for a universe that leads on to these glimmering mysteries and more beyond, perhaps to an infinity of budding universes. For Schleiermacher, the dogmas of Christianity and other religions were only metaphorical expressions of this fundamental feeling, this awareness of the power yet insufficiency of the human mind. This is not Gould's doctrine of separate magisteria, rather this view of religion is not merely compatible with science, it is necessary for the advancement of science. And, perhaps, for leading a coherent life, one in which the appreciation of poetry, art, and religion provide the same kind of experience that leads creative scientists to advance beyond their more pedestrian colleagues. Darwin was one such as these.

NOTES

CHARLES DARWIN: GREAT BRITON

1. Richard Dawkins, *A River Out of Eden* (New York: Basic Books, 1995), 133.

2. For background, I rely heavily on Linda Colley, *Britons: Forging the Nation, 1707–1837* (New Haven, CT: Yale University Press, 1992); Eric J. Evans, *The Forging of the Modern State: Early Industrial Britain, 1783–1870*, 3rd ed. (Harlow, Essex: Longman, 2001); and Boyd Hilton, *A Mad, Bad and Dangerous People? England, 1783–1846*, The New Oxford History of England (Oxford: Oxford University Press, 2006). M. Jonathan S. Hodge, "Capitalist Contexts for Darwinian Theory: Land, Finance, Industry and Empire," *Journal of the History of Biology* 42 (2009): 399–416, is very stimulating.

3. Robert Boyle, *A Free Enquiry into the Vulgarly Received Notion of Nature*, ed. Edward B. Davis and Michael Hunter (Cambridge: Cambridge University Press, 1996), 12.

4. Martin J. S. Rudwick, *Bursting the Limits of Time* (Chicago: University of Chicago Press, 2005); Michael Ruse, *The Gaia Hypothesis: Science on a Pagan Planet* (Chicago: University of Chicago Press, 2013).

5. Otto Mayr, *Authority, Liberty, and Automatic Machinery in Early Modern Europe* (Baltimore, MD: Johns Hopkins University Press, 1989).

6. Adam Smith, *An Inquiry into the Nature and Causes of the Wealth of Nations*, 2 vols. (London: W. Strahan and T. Cadell, 1776), 1:5.

7. Ibid., 1:22.

8. Ibid., 2:35.

9. Ibid., 1:17.

10. Thomas Robert Malthus, *An Essay on the Principle of Population* (London: Printed for J. Johnson, In St. Paul's Church-Yard, 1798; reprint, New York, Macmillan, 1996).

11. Ibid., chap. 3.

12. I take these figures from Evans, *Forging*, app. E.

13. Ibid., 163.

14. William Paley, *View of the Evidences of Christianity* (London: Faulder, 1794).

15. Michael Ruse, *Darwin and Design* (Cambridge, MA: Harvard University Press, 2003).

16. William Paley, *Natural Theology; or, Evidences of the Existence and Attributes of the Deity* (London: Faulder, 1802).

17. Michael Ruse, *Monad to Man: The Concept of Progress in Evolutionary Biology* (Cambridge, MA: Harvard University Press, 1996); *The Evolution-Creation Struggle* (Cambridge, MA: Harvard University Press, 2005).

18. I follow convention, using "Progress" (with a capital) for social progress, and "progress" (without a capital) for biological progress. Sometimes, it is difficult or impossible to distinguish the two, but that is precisely my point.

19. Thomas Robert Malthus, *Essay on a Principle of Population*, 6th ed. (London: John Murray, 1826), iv, xiv.

20. Erasmus Darwin, *The Temple of Nature* (London: J. Johnson, 1803), 1, lines 309–14.

21. Erasmus Darwin, *The Botanic Garden (Part I, The Economy of Vegetation)* (London: J. Johnson, 1791), 3, 349–50.

22. Erasmus Darwin, *Zoonomia; or, The Laws of Organic Life* (London: J. Johnson, 1794), 509.

23. Robert Chambers, *Vestiges of the Natural History of Creation* (London: Churchill, 1844).

24. This is from the third edition of *Vestiges* (1846), 400–402.

25. The definitive biography is Janet Browne, *Charles Darwin: Voyaging* (New York: Knopf, 1995), and *Charles Darwin: The Power of Place* (New York: Knopf, 2002). Michael Ruse, *The Cambridge Encyclopedia of Darwin and Evolutionary Thought* (Cambridge: Cambridge University Press, 2013), is very informative about Darwin's science and the culture within which he was immersed.

26. Adrian Desmond and James Moore, *Darwin's Sacred Cause: How a Hatred of Slavery Shaped Darwin's Views on Human Evolution* (New York: Houghton, Mifflin, Harcourt, 2009).

27. Henrietta Litchfield, *Emma Darwin: A Century of Family Letters* (Cambridge: Privately Printed, 1905).

28. Sam Schweber, "The Wider British Context of Darwin's Theorizing," in *The Darwinian Heritage*, ed. David Kohn (Princeton, NJ: Princeton University Press, 1985), 35–69.

29. W. Henry, *The Elements of Experimental Chemistry*, 8th ed. (London: Baldwin, Cradock and Joy, 1818), xix.

30. Samuel Parkes, *The Chemical Catechism, with Notes, Illustrations and Experiments* (London: Baldwin, Cradock and Joy, 1818), iii.

31. Adrian Desmond, "Robert E. Grant: The Social Predicament of a Pre-Darwinian Transmutationist," *Journal of the History of Biology* 17 (1984): 189–223.

32. Toby A. Appel, *The Cuvier-Geoffroy Debate: French Biology in the Decades Before Darwin* (Oxford: Oxford University Press, 1987).

33. Desmond and Moore, *Sacred*, 58.

34. William Kirby and William Spence, *An Introduction to Entomology; Or, Elements of the Natural History of Insects*, 2nd ed., 4 vols. (London: Longman, Hurst, Reece, Orme, and Brown, 1815–28).

35. Ibid., 1:xvi.

36. John S. Henslow, *Descriptive and Physiological Botany* (London: Longman, Rees, Orme, Brown, Green, and Longman; and John Taylor, 1836), is a short book based on the lectures.

37. John F. W. Herschel, *Preliminary Discourse on the Study of Natural Philosophy* (London: Longman, Rees, Orme, Brown, Green, and Longman, 1830).

38. Evans, *Forging*, app. D.

39. Charles Darwin, *Journal of Researches into the Natural History and Geology of the Countries Visited during the Voyage of H.M.S. Beagle round the World*, 2nd ed. (London: John Murray, 1845).

40. Charles Darwin, *The Autobiography of Charles Darwin, 1809–1882: With the Original Omissions Restored; Edited and with Appendix and Notes by His Grand-Daughter Nora Barlow* (London: Collins, 1958), 71.

41. Sandra Herbert, *Charles Darwin, Geologist* (Ithaca, NY: Cornell University Press, 2005).

42. Charles Lyell, *Principles of Geology: Being an Attempt to Explain the Former Changes in the Earth's Surface by Reference to Causes Now in Operation* (London: John Murray, 1830–33).

43. Georges Cuvier, *Essay on the Theory of the Earth*, trans. Robert Kerr (Edinburgh: W. Blackwood, 1813); Martin J. S. Rudwick, *The Meaning of Fossils* (New York: Science History Publications, 1972).

44. William Whewell, *History of the Inductive Sciences* (London: Parker, 1837), 3:588.

45. Adam Sedgwick, "Presidential Address to the Geological Society," *Proceedings of the Geological Society of London* 1 (1831): 281–316.

46. Martin J. S. Rudwick, "The Strategy of Lyell's *Principles of Geology*," *Isis* 61 (1969): 5–33.

47. This aspect of Lyell's theorizing is often known as "actualism."

48. James Hutton, *Theory of the Earth, With Proofs and Illustrations*, 2 vols. (Edinburgh: William Creech, 1795), 1:4.

49. Leonard G. Wilson, *Charles Lyell, the Years to 1841: The Revolution in Geology* (New Haven, CT: Yale University Press, 1972).

50. Rachel Laudan, *From Mineralogy to Geology: The Foundations of a Science, 1650–1830* (Chicago: University of Chicago Press, 1987).

51. Rudwick, *Bursting*; Ruse, *Gaia*.

52. Darwin, *Autobiography*, 86.

53. Leonard G. Wilson, ed., *Sir Charles Lyell's Scientific Journals on the Species Question* (New Haven, CT: Yale University Press, 1970).

54. Darwin, *Autobiography*, 91.

55. Baden Powell, *Essays on the Spirit of the Inductive Philosophy* (London: Longman, Brown, Green, and Longmans, 1855), 272.

56. Walter F. Cannon, "The Impact of Uniformitarianism: Two Letters from John Herschel to Charles Lyell, 1836–1837," *Proceedings of the American Philosophical Society* 105 (1961): 301–14.

57. Charles Babbage, *The Ninth Bridgewater Treatise: A Fragment* (London: John Murray, 1838).

58. Michael Ruse, *The Darwinian Revolution: Science Red in Tooth and Claw* (Chicago: University of Chicago Press, 1979); James Secord, *Victorian Sensation: The Extraordinary Publication, Reception, and Secret Authorship of "Vestiges of the Natural History of Creation"* (Chicago: University of Chicago Press, 2000).

59. Adam Sedgwick, "Vestiges," *Edinburgh Review* 82 (1845): 1–85; *Discourse on the Studies at the University of Cambridge*, 5th ed. (Cambridge: Cambridge University Press, 1850).

60. Mario Di Gregorio and N. W. Gill, eds., *Charles Darwin's Marginalia* (New York: Garland, 1990).

61. Paul H. Barrett, Peter J. Gautrey, Sandra Herbert, David Kohn, and Sydney Smith, eds., *Charles Darwin's Notebooks, 1836–1844* (Ithaca, NY: Cornell University Press, 1987), B: 101–2.

62. Michael Ruse, "Kant and Evolution," in *Theories of Generation*, ed. J. Smith (Cambridge: Cambridge University Press, 2006), 402–15.

63. Michael Ruse, "Charles Darwin and Artificial Selection," *Journal of the History of Ideas* 36 (1975): 339–50.

64. John Sebright, *The Art of Improving the Breeds of Domestic Animals in a Letter Addressed to the Right Hon. Sir Joseph Banks, K.B.* (London: John Harding, 1809).

65. Barrett et al., *Notebooks*, D: 135e.

66. Barrett et al., *Notebooks*, N: 42.

67. William Whewell, *Philosophy of the Inductive Sciences*, 2 vols. (London: Parker, 1840).

68. John F. W. Herschel, "Review of Whewell's History and Philosophy," *Quarterly Review* 135 (1841): 177–238; Michael Ruse, "Darwin's Debt to Philosophy: An Examination of the Influence of the Philosophical Ideas of John F. W. Herschel and William Whewell on the Development of Charles Darwin's Theory of Evolution," *Studies in History and Philosophy of Science* 6 (1975): 159–81.

69. Whewell, *Philosophy*, 2:230.

70. John F. W. Herschel, "Light," in *Encyclopedia Metropolitana*, ed. Edward Smedley et al. (London: J. Griffin, 1827).

71. Whewell, *Philosophy*, 1:303.

72. Charles Darwin, *The Variation of Animals and Plants under Domestication*, 2 vols. (London: John Murray, 1868).

73. Charles Darwin and Alfred Russel Wallace, *Evolution by Natural Selection*, ed.

Gavin de Beer (Cambridge: Cambridge University Press, 1958). The earlier essay (1842) is known (by Darwin scholars) as the "Sketch," and the later (1844) as the "Essay."

74. Michael Ruse, "Natural Selection in the Origin of Species," *Studies in History and Philosophy of Science* 1 (1971): 311–52.

75. Darwin and Wallace, *Evolution*, 116 (from the "Essay").

76. Darwin and Wallace, *Evolution*, 119 (from the "Essay").

77. Darwin and Wallace, *Evolution*, 229–30 (from the "Essay").

78. Jack Morrell and Arnold Thackray, *Gentlemen of Science: Early Years of the British Association for the Advancement of Science* (Oxford: Oxford University Press, 1981).

79. Mike Dixon and Gregory Radick, *Darwin in Ilkley* (Ikley: History Press, 2009).

80. Charles Darwin, *On the Origin of Species by Means of Natural Selection; or, The Preservation of Favoured Races in the Struggle for Life* (London: John Murray, 1859).

81. Ibid., 84.

82. Michael Ruse, "Darwin and Mechanism: Metaphor in Science," *Studies in History and Philosophy of Biology and Biomedical Sciences* 36 (2005): 285–302.

83. Richard Owen, *On the Archetype and Homologies of the Vertebrate Skeleton* (London: Voorst, 1848); *On the Nature of Limbs* (London: Voorst, 1849).

84. Thomas Henry Huxley, "On the Morphology of the Cephalus Mullusca, as Illustrated by the Anatomy of Certain Heteropoda and Pteropoda Collected during the Voyage of the HMS Rattlesnake in 1846–50," *Philosophical Transactions of the Royal Society* 143, no. 1 (1853): cxliii, 1, 29–66.

85. William Whewell, *The Plurality of Worlds* (London: Parker, 1853).

86. Charles Darwin, *A Monograph of the Fossil Lepadidae; or, Pedunculated Cirripedes of Great Britain* (London: Paleontographical Society, 1851); *A Monograph of the Sub-Class Cirripedia, with Figures of All the Species. The Lepadidae; or Pedunculated Cirripedes* (London: Ray Society, 1851); *A Monograph of the Fossil Balanidae and Verrucidae of Great Britain* (London: Paleontographical Society, 1854); *A Monograph of the Sub-Class Cirripedia, with Figures of All the Species. The Balanidge (or Sessile Cirripedes); the Verrucidae, and C.* (London: Ray Society, 1854).

87. Darwin, *Origin*, 206.

88. Darwin and Wallace, *Evolution*, 114 (from the "Essay").

89. Charles Darwin, *Origin of Species*, 3rd ed. (London: John Murray, 1861), 85.

90. Ruse, *Darwin and Design*.

91. Asa Gray, *Darwiniana* (New York: Appleton, 1876).

92. Letter from Darwin to Asa Gray, May 22, 1860, in Charles Darwin, *The Collected Correspondence of Charles Darwin*, 21 vols. to date (Cambridge: Cambridge University Press, 1985–), 8:224.

93. I am entirely comfortable with the suggestion that Darwin got the necessity for

keeping God out of science from Whewell and that Whewell got this from Kant; although, as far back (at least) as Robert Boyle toward the end of the seventeenth century we find the instruction to separate science (where God is not allowed) from natural theology (where God is the chief player). See Ruse, *Darwin and Design*.

94. Hilton, *Mad, Bad*, 304.
95. Ruse, *Gaia*, 93.
96. Smith, *Wealth*, 1:1.
97. Sam Schweber, "Darwin and the Political Economists: Divergence of Character," *Journal of the History of Biology* 13 (1980): 195–289.
98. Darwin, *Origin*, 93–94.
99. Ibid., 94.
100. Ibid., 115–16.
101. It is true that Darwin's immediate influence was the French zoologist Henri Milne-Edwards. But Milne-Edwards, whose father incidentally was British, obviously got it from earlier sources, and Darwin likewise was exposed to these sources. My sense is that Darwin did not refer directly to Smith because he wanted his work to be taken as a work of professional science and not a popular work in the style of *Vestiges*. I am not saying he was entirely successful in this and would myself agree that there is a lot more natural theology hovering than is acceptable in professional science today or even back then.
102. *Origin*, 63–64.
103. Ibid., 75.
104. Michael Ruse, "Charles Darwin and Group Selection," *Annals of Science* 37 (1980): 615–30.
105. William Hamilton, "The Genetical Evolution of Social Behaviour," *Journal of Theoretical Biology* 7 (1964): 1–52.
106. I do not think that David Hume was an evolutionist, and he certainly knew nothing of modern genetics, but there is a remarkable anticipation in Hume's *Treatise* of the idea behind kin selection. Hume writes: "A man naturally loves his children better than his nephews, his nephews better than his cousins, his cousins better than strangers, where every thing else is equal. Hence arise our common measures of duty, in preferring the one to the other. Our sense of duty always follows the common and natural course of our passions." David Hume, *A Treatise of Human Nature*, ed. D. J. Norton and M. J. Norton (Oxford: Oxford University Press, 2000), 311.
107. Darwin, *Origin*, 236.
108. Ibid., 237–38.
109. Ibid., 238. In all six editions of the *Origin*, when Darwin uses the word "community" referring to a group of organisms, he is referring to a nest or hive of hymenoptera—ants, bees, and wasps.

110. Ruse, *Monad*.

111. Barrett et al., *Notebooks*, B: 74.

112. Ibid., B: 207.

113. Ibid., E: 95–96.

114. George F. W. Hegel, *Philosophy of Nature* (Oxford: Oxford University Press, [1817] 1970), 21.

115. Letter from Darwin to Charles Lyell, October 11, 1859, in Darwin, *Correspondence*, 7:343.

116. Darwin and Wallace, *Evolution*, 42 (from the "Sketch").

117. Darwin, *Origin*, 345.

118. Ibid., 338.

119. Ibid.

120. Darwin, *Origin*, 3rd ed., 134.

121. Darwin and Wallace, *Evolution*, 87 (from the "Sketch").

122. David Brewster, "Review of Comte's 'Cours de Philosophie Positive,'" *Edinburgh Review* 67 (1838): 271–308. My italics flag words taken up by Darwin.

123. Charles Darwin, *On the Various Contrivances by Which British and Foreign Orchids Are Fertilized by Insects, and On the Good Effects of Intercrossing* (London: John Murray, 1862).

124. Henslow, *Botany*, 44.

125. Charles Darwin, *The Descent of Man and Selection in Relation to Sex*, 2 vols. (London: John Murray, 1871) and *The Expression of Emotions in Man and in Animals* (London: John Murray, 1872).

126. Thomas Henry Huxley, *Evidence as to Man's Place in Nature* (London: Williams and Norgate, 1863).

127. Alfred Russel Wallace, "The Origin of Human Races and the Antiquity of Man Deduced from the Theory of Natural Selection," *Journal of the Anthropological Society of London* 2 (1864): clvii–clxxxvii.

128. Alfred Russel Wallace, *Contributions to the Theory of Natural Selection: A Series of Essays* (London: Macmillan, 1870).

129. Alfred Russel Wallace, "Sir Charles Lyell on Geological Climates and the Origin of Species," *Quarterly Review* 126 (1869): 359–94, 392.

130. Letter from Darwin to Alfred Russel Wallace, March 27, 1869, Darwin, *Correspondence*, 17:157.

131. Darwin, *Origin*, 488.

132. Darwin, *Descent*, 1:238.

133. Ibid., 1:174–75.

134. Ibid., 2:315.

135. Ibid., 1:169.

136. Ibid., 2:368–69.

137. Ibid., 1:316.

138. Ibid., 1:320.

139. Charles Darwin, *The Descent of Man*, 2nd ed. (London: John Murray, 1874), 259.

140. Ibid., 259–60.

141. Hume, *Religion*, 78.

142. Darwin, *Descent*, 1:67.

143. Frederick W. H. Myers, "George Eliot," *Century Magazine*, November 1881.

144. Letter from Darwin to Heinrich Fick, July 26, 1872, in Darwin, *Correspondence*, 20:324.

145. Darwin, *Descent*, 1:80.

146. Ibid., 1:71–72.

147. There is no reason to think that Darwin read the *Treatise* or flagged that remarkable anticipation of kin selection. The point rather is that the tenor of Darwin's thinking belonged to the school of which Hume was the most distinguished member.

148. Ibid., 1:86. There are differences between Hume and Smith, with the former more inclined to think we have the same feelings as others and the latter giving a greater role for the imagination and thus allowing that we might care about someone even though we don't share their exact feelings—as, let us say, we might regard a madman. Hume is more inclined to put things down to nature whereas for Smith socialization is crucial. These are subtleties we can gloss over here, although philosophically speaking the Hume-Smith differences may be significant—Smith seems to open morality more to critical judgment and hence in this sense is closer to moral realism than Hume. If one were forced to judge, I would put Darwin closer to Hume than Smith, but the argument could go the other way.

149. Ibid., 1:169.

150. Ibid., 1:163.

151. Robert Trivers, "The Evolution of Reciprocal Altruism," *Quarterly Review of Biology* 46 (1971): 35–57.

152. Darwin, *Descent*, 1:163–64.

153. Ibid., 1:164.

154. Ibid., 1:165.

155. Ibid. 1:166.

156. Wallace, "Origin," clxii.

157. Darwin, *Descent*, 1:85.

158. Ibid., 1:66.

159. Ibid., 1:159.

160. Ibid., 1:161.

161. Henry Sidgwick, "The Theory of Evolution in Its Application to Practice," *Mind* 1 (1876): 52–67.

162. Unpublished letter from Darwin to George Darwin, April 27, 1876. *Darwin Correspondence Project*, letter 10478.

163. Darwin, *Descent*, 1:73–74.

164. Almost certainly exaggerated, but with a grain of truth nevertheless. See John R. Lucas, "Wilberforce and Huxley: A Legendary Encounter," *Historical Journal* 22 (1979): 313–30.

165. E. Lurie, *Louis Agassiz: A Life in Science* (Chicago: University of Chicago Press, 1960).

166. Charles Hodge, *What Is Darwinism?* (New York: Scribner's, 1874).

167. He didn't say that, but that was the message heard and accepted.

168. M. Artigas, Thomas F. Glick, and R. A. Martinez, *Negotiating Darwin: The Vatican Confronts Evolution, 1877–1902* (Baltimore, MD: Johns Hopkins University Press, 2006).

169. Benjamin Disraeli, *Lothair*, 2 vols. (Leipzig: Bernhard Tauchnitz, 1870), 1:193: "Instead of Adam, our ancestry is traced to the most grotesque of creatures, thought is phosphorus, the soul complex nerves, and our moral sense a secretion of sugar. Do you want these views in England?"

170. I refer of course to *Call of Wild*, first published in 1903. To be fair, it is a terrific story. I am not surprised—although very envious—to find that there are over thirty (!) editions still in print. This underlines and extends to America the points I am making here. In truth, London was far more sophisticated than the usual crude Social Darwinian caricature. See Lawrence I. Berkove, "Jack London and Evolution: From Spencer to Huxley," *American Literary Realism* 36 (2004): 243–55.

171. Apparently the wife "could not undertake the volumes of Herbert Spencer." I know how she felt. Fabio Cleto, "The Biological Drama: Darwinian Ethics in George Gissing's Fiction," *Gissing Journal* 28, no. 3 (1992): 1–13, makes some illuminating comments on Gissing's use of Darwinism.

172. Peter Bowler, *The Non-Darwinian Revolution: Reinterpreting a Historical Myth* (Baltimore, MD: Johns Hopkins University Press, 1988); Michael Ruse, "The Darwinian Revolution: Rethinking Its Meaning and Significance," *Proceedings of the National Academy of Sciences* 106 (2009): 10040–7. Note my qualifications in the text. If we take account of public reactions, speaking simply of the Darwinian Revolution as a "myth" is nigh-willful nonsense. And that is before you start to take into account the massive influence of Darwinian themes on American pragmatism. See Michael Ruse, ed., *Philosophy after Darwin: Classic and Contemporary Themes* (Princeton, NJ: Princeton University Press, 2009). Also Michael Ruse, *Darwinism as Religion: What Literature Tells Us about Evolution* (New York: Oxford University Press, 2016).

173. Joseph D. Burchfield, *Lord Kelvin and the Age of the Earth* (New York: Science History Publications, 1975).

174. Gray, *Darwiniana*.

175. Robert J. Richards, *The Tragic Sense of Life: Ernst Haeckel and the Struggle over Evolutionary Thought* (Chicago: University of Chicago Press, 2008).

176. John Holmes, "Literature and Science vs. History of Science," *Journal of Literature and Science* 5, no. 2 (2012): 67–71.

177. Julian Huxley, *Evolution: The Modern Synthesis* (London: Allen and Unwin, 1942).

178. I put in this qualification. Whatever he thought of the utility of natural selection in professional science, when it came to his more popular writing, as in his *Evolution and Ethics* (1893), Huxley was as open and accepting of natural selection as any poet or novelist.

179. William Provine, *The Origins of Theoretical Population Genetics* (Chicago: University of Chicago Press, 1971).

180. Think of the hatchet job done by Lytton Strachey in his *Eminent Victorians*, published in 1918.

181. Michael Ruse, *The Evolution Wars*, 2nd ed. (Millerton, NY: Greyhouse Publishing, 2009).

182. Richard Dawkins, *The Selfish Gene* (Oxford: Oxford University Press, 1976); the Margaret Thatcher quote is from an interview she gave to *Woman's Own* magazine, September 23, 1987, while she was still prime minister.

183. Edward O. Wilson, *Sociobiology: The New Synthesis* (Cambridge, MA: Harvard University Press, 1975). The classic attack, with dozens of signatures, including Richard Lewontin, Richard Levins, and Stephen Jay Gould, not just at Harvard like Wilson but in the same department, was Elizabeth Allen et al., "Letter to the Editor," *New York Review of Books* 22, no. 18 (1975): 43–44. The controversy is discussed in Michael Ruse, *Sociobiology: Sense or Nonsense* (Dordrecht: Reidel, 1979).

184. Ruse, *Gaia*, 175.

185. Letter from Darwin to Heinrich Frick, July 26, 1872, in Darwin, *Correspondence*, 20:324.

186. Richard Weikardt, "A Recently Discovered Darwin Letter on Social Darwinism," *Isis* 86 (1995): 609–11.

187. Michael Ruse, *Gaia Hypothesis*; "Adaptive Landscapes and Dynamic Equilibrium: The Spencerian Contribution to Twentieth-Century American Evolutionary Biology," in *Darwinian Heresies*, ed. Abigail Lustig, Robert J. Richards, and Michael Ruse (Cambridge: Cambridge University Press, 2004), 131–50.

188. Stephen Jay Gould and Richard Lewontin, "The Spandrels of San Marco and the Panglossian Paradigm: A Critique of the Adaptationist Programme," *Proceedings of the Royal Society of London, Series B: Biological Sciences* 205 (1979): 581–98, 581.

189. Elliott Sober, *Did Darwin Write the "Origin" Backwards?* (Buffalo, NY: Prometheus Books, 2010).

190. Janet Browne, *Voyaging*, xiii.

CHARLES DARWIN: COSMOPOLITAN THINKER

1. George Bernard Shaw, preface to *Back to Methuselah* (London: Penguin Books, [1921] 1961), 44.

2. Stephen Jay Gould, *The Structure of Evolutionary Theory* (Cambridge, MA: Harvard University Press, 2002), 122. Gould recognized (119), as have most scholars, that Darwin was attracted to Paley in his youth and that even certain structural aspects of the *Origin* sound a bit like Paley's *Natural Theology*, for instance, in going from artificial contrivance (a watch found on the beach) to the natural situation (animals as contrived machines).

3. Daniel Dennett, *Darwin's Dangerous Idea* (New York: Simon and Schuster, 1995), 133.

4. Michael Ruse, *Darwin and Design* (Cambridge, MA: Harvard University Press, 2003), 122.

5. Elliott Sober does recognize that in the *Origin*, Darwin did make explicit appeal to God as the promulgator of natural law. Sober, however, attempts to parry aside this feature of the *Origin* by claiming this was part of Darwin's philosophical persuasion and not part of his scientific analysis. Sober quite obviously attempts to sanitize Darwin's science, rendering it acceptable to contemporary biologists. See Elliott Sober, "Darwin and Naturalism," in his *Did Darwin Write the Origin Backwards? Philosophical Essays on Darwin's Theory* (Amherst, NY: Prometheus Books, 2011), 121–52. I will consider Sober's positions in more detail below.

6. Darwin to Asa Gray (May 22, 1860), in *The Correspondence of Charles Darwin*, ed. Frederick Burkhardt et al., 21 vols. to date (Cambridge: Cambridge University Press, 1985–), 8:224.

7. Charles Darwin, *The Autobiography of Charles Darwin*, ed. Nora Barlow (New York: Norton, 1958), 93–94.

8. See, for example, Phillip Sloan, "Darwin, Vital Matter, and the Transformation of Species," *Journal of the History of Biology* 19 (Fall 1986): 369–445; and "'The Sense of Sublimity': Darwin on Nature and Divinity," *Osiris*, 2nd series, 16 (2001): 251–69.

9. See, for example, John Hedley Brooke, "'Laws Impressed on Matter by the Creator'? The Origin and the Question of Religion," in *The Cambridge Companion to the "Origin of Species,"* ed. Michael Ruse and Robert J. Richards (Cambridge: Cambridge University Press, 2009), 256–74.

10. See, for example, John Cornell, "Newton of the Grassblade? Darwin and the Problem of Organic Teleology," *Isis* 77 (1986): 404–21; and "God's Magnificent Law: The Bad Influence of Theistic Metaphysics on Darwin's Estimation of Natural Selection," *Journal of the History of Biology* 20 (1987): 381–412.

11. See, for example, essays in Robert J. Richards, *Was Hitler a Darwinian? Disputed*

Questions in the History of Evolutionary Theory (Chicago: University of Chicago Press, 2013).

12. R. C. Lewontin, Steven Rose, and Leon Kamin, *Not in Our Genes* (New York: Pantheon, 1984), 51.

13. The French should not be forgotten: the likes of George Cuvier, Etienne and Geoffroy Saint-Hilaire, and Jean Baptiste de Lamarck among others were also influential. Darwin was a cosmopolitan thinker. I do, however, believe that German thought has been extremely influential on Darwin's conceptions and that this strand has been either neglected or denied.

14. Erasmus Darwin's *Botanic Garden* comprises two poems, initially published anonymously and separately as *The Loves of the Plant* (1789) and *The Economy of Vegetation* (1791). Both appeared together posthumously under Darwin's name in 1825: *The Botanic Garden: A Poem in Two Parts* (London: Jones and Co., 1825).

15. Samuel Taylor Coleridge, *Biographia Literaria*, vol. 7, *The Collected Works of Samuel Taylor Coleridge*, ed. James Engell and W. Jackson Bate (Princeton, NJ: Princeton University Press, 1983), 19. Coleridge thought Darwin's poetry insipid, the kind an industrious milliner might fabricate. Though Goethe shared Coleridge's judgment about the aesthetic value of Darwin's poetry, the *Botanic Garden* may well have suggested to him that he might produce a long poem expressive of his scientific ideas. See Robert J. Richards, *The Romantic Conception of Life: Science and Philosophy in the Age of Goethe* (Chicago: University of Chicago Press, 2002), 466.

16. I have discussed the impact of Erasmus Darwin's *Zoonomia* on German Romantic thought in ibid., 300-306, 466-67.

17. Charles Darwin, *The Autobiography of Charles Darwin*, ed. Nora Barlow (New York: Norton, 1958), 28.

18. Ibid., 49-51.

19. For a discussion of Darwin's introduction to German transcendental biology, see Phillip Sloan, "Darwin, Vital Matter, and the Transformation of Species," *Journal of the History of Biology* 19 (Fall 1986): 369-445.

20. Darwin, *Autobiography*, 59.

21. Alexander von Humboldt and Aimé Bonpland, *Personal Narrative of Travels to the Equinoctial Regions of the New Continent, during the Years 1799-1804*, trans. Helen Williams, 7 vols. (London: Longman, Hurst, Rees, Orme, and Brown, 1818-29). Though Bonpland's name is listed on the title page, the narrative is in the first person, Humboldt being that person.

22. Darwin to Joseph Hooker (February 10, 1845), in *Correspondence of Charles Darwin*, 3:140.

23. Darwin to his sister Caroline (April 25-26, 1832), in *Correspondence of Charles Darwin*, 1:226.

24. Charles Darwin, *Darwin's Ornithological Notes*, ed. Nora Barlow, *Bulletin of the British Museum (Natural History)*, Historical series, 2, no. 7 (1963): 262.

25. Frank Sulloway discusses the role of the mockingbirds in Darwin's discovery in "Darwin's Conversion: The *Beagle* Voyage and Its Aftermath," *Journal of the History of Biology* 15 (1982): 325–96.

26. Darwin, *Red Notebook* (MS, 127–33), in *Charles Darwin's Notebooks, 1836–1844*, ed. Paul Barrett et al. (Ithaca, NY: Cornell University Press, 1987), 61–63.

27. These several notebooks can be found in ibid.

28. These essays may be found in *Foundations of the Origin of Species: Two Essays Written in 1842 and 1844*, ed. Francis Darwin (Cambridge: Cambridge University Press, 1909).

29. Charles Darwin and Richard Owens, eds., *The Zoology of the Voyage of H.M.S. Beagle, under the Command of Captain FitzRoy, during the Years 1832 to 1836*, 5 parts (London: Smith Elder and Co., 1838–43).

30. Charles Darwin, *The Structure and Distribution of Coral Reefs. Being the First Part of the Geology of the Voyage of the Beagle* (London: Smith Elder and Co., 1842); *Geological Observations on the Volcanic Islands Visited during the Voyage of H.M.S. Beagle, Together with Some Brief Notices of the Geology of Australia and the Cape of Good Hope. Being the Second Part of the Geology of the Voyage of the Beagle* (London: Smith Elder and Co., 1844); *Geological Observations on South America. Being the Third Part of the Geology of the Voyage of the Beagle* (London: Smith Elder and Co., 1846). Sandra Herbert gives a superb account of Darwin's geological researches in her *Charles Darwin, Geologist* (Ithaca, NY: Cornell University Press, 2005).

31. Charles Darwin, *Living Cirripedia, A Monograph on the Sub-Class Cirripedia, with Figures of All the Species*, vol. 1, *The Lepadidæ; or, Pedunculated Cirripedes* (London: The Ray Society, 1852); *Living Cirripedia*, vol. 2, *The Balanidæ (or Sessile Cirripedes); the Verrucidæ* (London: The Ray Society, 1854); *A Monograph on the Fossil Lepadidae; or, Pedunculated Cirripedes of Great Britain* (London: Printed for the Palæontographical Society, 1851); *A Monograph on the Fossil Balanidae and Verrucidae of Great Britain* (London: Printed for the Palæontographical Society, 1854).

32. Darwin, *Notebook B* (MS, 1), in *Charles Darwin's Notebooks*, 170.

33. Charles Darwin, Personal Journal, MS, 34, Cambridge University Library, DAR 158.1-76.

34. Wallace's letter is missing. Darwin described it to Charles Lyell in a letter (June 18, 1858), in *Correspondence of Charles Darwin*, 107.

35. Ibid.

36. The sixth edition of the *Origin* was reprinted in 1876 with some small number of additions and corrections.

37. This was Haeckel's recollection of a meeting with Darwin at his home in the village of Downe. Haeckel was on his way to the Canary Islands, but stopped in London to see English biologists, especially Darwin. Darwin invited Haeckel to visit him on Sunday, October 21, 1866. The recollection was recorded by

Haeckel's disciple Wilhelm Bölsche, in *Ernst Haeckel: Ein Lebensbild* (Berlin: Georg Bondi, 1909), 179.

38. Charles Darwin, *On the Various Contrivances by Which British and Foreign Orchids Are Fertilised by Insects* (London: John Murray, 1862); "On the Movements and Habits of Climbing Plants," *Journal of the Linnaean Society of London (Botany)* 9 (1865): 1–118 (later published as a small book); *The Variation of Animals and Plants under Domestication*, 2 vols. (London: Murray, 1868).

39. See Darwin's discussion of Galton's experiments in his essay "Pangenesis," *Nature* 3 (April 27): 502–503.

40. Charles Darwin, *The Descent of Man, and Selection in Relation to Sex*, 2 vols. (London: Murray, 1871). A second edition was published in 1874.

41. Phillip Prodger explores Darwin's use of photography in *Expression of the Emotions*, especially the photography of Oscar Rejlander and Guillaume-Benjamin Duchenne de Boulogne. See Phillip Prodger, *Darwin's Camera: Art and Photography in the Theory of Evolution* (Oxford: Oxford University Press, 2009).

42. I have discussed the character of Darwin's argument in Robert J. Richards, *Darwin and the Emergence of Evolutionary Theory of Mind and Behavior* (Chicago: University of Chicago Press, 1987), 230–34.

43. Charles Darwin, *Insectivorous Plants* (London: Murray, 1875; 2nd ed., 1888); *The Effects of Cross and Self Fertilisation in the Vegetable Kingdom* (London: Murray, 1876; 2nd ed., 1878); *The Different Forms of Flowers on Plants of the Same Species* (London: Murray, 1877); *The Power of Movement in Plants* (London: Murray, 1880); and *The Formation of Vegetable Mould, through the Action of Worms* (London: Murray, 1881).

44. Charles Lyell, *The Principles of Geology*, 3 vols. (London: Murray, 1830–33).

45. Darwin, *Autobiography*, 67–68.

46. Goethe in conversation with Eckermann (December 11, 1826). See Johann Peter Eckermann, *Gespräche mit Goethe in den letzten Jahren seines Lebens*, 3rd ed. (Berlin: Aufau-Verlag, 1987), 161.

47. See Georg Forster, *A Voyage Round the World in His Britannic Majesty's Sloop, Resolution, Commanded by Capt. James Cook, during the Years 1772, 3, 4, and 5*, 2 vols. (London: B. White, 1777). The book was also in Darwin's library. See *Correspondence of Charles Darwin*, 2:222.

48. If one measured from the center of the earth, Chimborazo—because of the equatorial bulge of the earth and the plateau—is the highest mountain at twenty thousand feet above sea level.

49. See Laura Dassow Walls, *The Passage to Cosmos: Alexander von Humboldt and the Shaping of America* (Chicago: University of Chicago Press, 2009), 13–14.

50. Alexander von Humboldt, *Humboldt über das Universum: Alexander von Humboldt, Die Kosmosvorträe 1827/28* (Frankfurt a.M.: Insel Verlag, 1993).

51. Darwin at least read the first of a two-volume set in the English translation:

Alexander von Humboldt, *Cosmos*, trans. E. Sabine, 2 vols. (London: Longman, Brown, Green & Longmans, 1846).

52. Darwin, *Autobiography*, 68.

53. Humboldt and Bonpland, *Personal Narrative*, 1:xlv–xlvi.

54. In addition to the *Personal Narrative*, the four other books of Humboldt that Darwin had with him were *Essai géognostique sur le gisement des roches dans les deux hemispheres; Fragmens de géologie et de climatologie asiatiques* (2 vols.); *Political Essay on the Kingdom of New Spain*; and *Tableaux de la nature* (French translation of 2nd ed. of *Anschiten der Natur*, 2 vols.). See the list of books that Darwin had on the *Beagle* in appendix iv of *Correspondence of Charles Darwin*, 1:553–66.

55. Charles Darwin, *Beagle Diary*, ed. R. D. Keynes (Cambridge: Cambridge University Press, 1988), 42 (entry for February 28, 1832).

56. For these subjects, see ibid., 34 (February 6, 1832), 42 (February 18, 1832), 48 (March 24–26, 1832), 54 (April 19, 1832), 67 (May 26, 1832), 70 (June 2, 1832), 72 (June 4, 1832), 267 (November 26, 1834), 288 (February 12, 1835), and 308 (March 21, 1835).

57. Caroline Darwin to Charles Darwin (October 28, 1833), in *Correspondence of Charles Darwin*, 1:345.

58. Alexander von Humboldt, *Cosmos: A Sketch of the Physical Description of the Universe*, trans. E. C. Otté, 5 vols. (New York: Harper, n.d.), 1:25.

59. Ibid., 2:80.

60. Prior to the publication of his evaluation of Darwin's abilities in *Kosmos*, Humboldt had sent Darwin a long letter praising him extravagantly for his *Journal of Researches* and suggesting that the young naturalist had great promise for further discovery. See Alexander von Humboldt to Darwin (September 18, 1839), in *Correspondence of Charles Darwin*, 2:218–22.

61. Darwin, *Beagle Diary*, 443. Darwin transferred this judgment to his published account of the voyage. See Charles Darwin, *Journal of Researches into the Geology and Natural History of the Various Countries Visited by H.M.S. Beagle* (London: Henry Coburn, 1839), 604.

62. Darwin, *Beagle Diary*, 444; *Journal of Researches*, 604.

63. Hugh Trevor-Roper, "The Romantic Movement and the Study of History," in his *History and the Enlightenment* (New Haven, CT: Yale University Press, 2010), 187.

64. Charles Darwin, *Notebook M* (MS, 34), in *Charles Darwin's Notebooks*, 527.

65. Darwin, *Notebook C* (MS, 210e–211), in *Charles Darwin's Notebooks*, 305.

66. Carl Gustav Carus, *Lebenserinnerungen und Denkwürdigkeiten*, 4 vols. (Leipzig: F. A. Brockhaus, 1865–66), 1:129.

67. Johann Wolfgang von Goethe, *Versuch die Metamorphose der Pflanzen zu erklären* (Gotha: Carl Wilhelm Ettinger, 1790).

68. Carl Gustav Carus, *Von den Ur-Theilen des Knochen- und Schalengerüstes* (Leipzig: Gerhard Fleischer, 1828), ix.

69. See Richard Owen, *On the Archetype and Homologies of the Vertebrate Skeleton* (London: John Van Voorst, 1848), 70. Owen mentions Goethe's contribution to the discussion of the vertebral theory of the skull and suggests the artistic merit of Carus's illustrations in *Von der Ur-Theilen*. He does not suggest his own illustration (fig. 16) owes anything to Carus. Rupke details Owen's reluctant and gradual admission of his debt to Carus. See Nicolass Rupke, *Richard Owen, Biology without Darwin*, 2nd ed. (Chicago: University of Chicago Press, 2009), 124–25.

70. Owen, *On the Archetype, 7*.

71. In his two books, published in rapid succession—*On the Archetype* (1848) and *On the Nature of Limbs* (1849)—Owen postulated vital forces that would produce a quasi-developmental evolution of the vertebrate species. But his configuration of these forces differed considerably from the first to the second book. In the prior book, he distinguished the Platonic ἰδέα, which produced the variety of species forms, from the "polarizing force," to which he ascribed the basic form of the vertebrates—the archetype (*On the Archetype*, 172). The notion of a polarizing force was part of the philosopher Friedrich Schelling's repertoire. A year later, Owen identified the archetype with the *idea*: the "predetermined pattern, answering to the 'idea' of the Archetypal World in the Platonic cosmogony" (*On the Nature of Limbs*, 2). Rupke believes that Owen's explicit Platonizing of the archetype, making it an idea, rendered it more obviously as a candidate for residing in the mind of God and moved his theory away from German *Naturphilosohie* (Rupke, 133). This interpretation, while it may be correct, faces two obstacles. The last paragraph of the first book ascribed the idea, which produced the diversity of species, as the means by which the divine mind adapted organisms to their specific environments; this interpretation suggested, according to Owen, "superior design, intelligence, and foresight" (172). This construal would seem to have been an adequate prophylactic against the German threat. Further, in his book *On the Nature of Limbs*, Owen specified that the word "nature" in the title should be understood as the same as the German *Bedeutung*, that is, "meaning"—thus making explicit a German linkage. Moreover, at the end of this second book (86), he contended that we ought to ascribe to Nature (as a secondary cause) the progressive development of animals. So Owen may have changed his way of understanding the archetype, but the second way seems no more distant from German *Naturphilosophie* than the first. I think the notion of a Platonic idea as the basic template—the archetype—was simply less confusing than his original formulation.

72. This inscription is on the back flyleaf of Darwin's copy of Richard Owen's *On the Nature of Limbs*; the book is in the Manuscript Room of Cambridge University Library.

73. Darwin, *Origin of Species*, 434–39. Unless otherwise specified, it is the first edition of Darwin's works to which the notes refer.

74. Carus, *Von den Ur-Theilen*, vii: "The history of the sciences, especially the natural sciences, offers a rich load of comparative observations. One thinks of the
common deployment of the idea of a parallelism between the [embryological]
development [*Entwickelung*] of the more advanced animal forms—yes, even
man himself—and the particular classes and species within the animal kingdom."

75. Charles Darwin, *Notebook B* (MS, 1), in *Charles Darwin's Notebooks, 1836–1844*,
ed. Paul Barrett et al. (Ithaca, NY: Cornell University Press, 1987), 170.

76. Erasmus Darwin, *Zoonomia; or, The Laws of Organic Life*, 2 vols. (London: J. Johnson, 1794–96), 1:505.

77. Darwin differed from his grandfather in the explanation of individual development. Erasmus Darwin held that the embryo—the unformed embryo—in
the human received propensities for development from the imagination of the
male operating on the seminal fluid. These propensities, accounting for similarities to the parent, would be expressed in the maternal womb through the
habits of the developing embryo (E. Darwin, *Zoonomia*, 1:409).

78. Darwin, *Origin of Species*, 450.

79. Henry Crabb Robinson, *Diary, Reminiscences, and Correspondence*, ed. Thomas
Sadler, 3rd ed., 2 vols. (London: Macmillan and Co., 1872), 2:226. The same
anecdote was quoted by Harriet Martineau in *Chamber's Encyclopedia of English
Literature*, ed. David Patrick, 2nd ed., 3 vols. (London: W. R. Chambers, 1903),
3:389.

80. Darwin, *Autobiography*, 66.

81. Laura Snyder, *Reforming Philosophy* (Chicago: University of Chicago Press,
2006), 43n.

82. William Whewell, *History of the Inductive Sciences, from Earliest to Present Times*,
3rd ed., 3 vols. (London: Parker & Son, [1837] 1857): 3:478. The changes in the
subsequent editions are marked in separate sections. The main text is that of
the first edition of 1837.

83. Ibid., 476.

84. Darwin, "Essay of 1842," in *Foundations of the Origin of Species*, 23.

85. Ibid., 57.

86. Ibid., 92–93.

87. Darwin, "Essay of 1842," in *Foundations of the Origin of Species*, 51. Comparable
remarks are found in the "Essay of 1844," in ibid., 253–54, and the *Origin of
Species*, 485. Square brackets indicate an erasure. The remark about it being beneath the dignity of the Creator to have produced slimy worms echoes a passage
in Plato that Darwin may have read as part of his classics training at Cambridge.
In the *Parmenides*, the figure of Socrates doubts there could be Forms of such
undignified objects as hair, mud, and dirt. See Plato, *Parmenides*, 130c-d.

88. William Whewell, *Astronomy and General Physics Considered with Reference to
Natural Theology* (Bridgewater Treatise) (Philadelphia: Carey, Lea & Blanchard,

1833), 267. Darwin could have just as well quoted from an author who had special appeal for the young scholar and an individual he visited on the return voyage to England, John Herschel. In his *Preliminary Discourse on the Study of Natural Philosophy* (London: Longman, Rees, Orme, Brown, and Green, 1831), Herschel confirmed that the deity created through general laws, but did not specify every contingency in nature (37).

89. Darwin, *Origin of Species*, 469, 475, 482, and 488.

90. Ibid., 488.

91. Darwin to Asa Gray (May 22, 1860), in *Correspondence of Charles Darwin*, 8:224. Darwin seemed here to be following Herschel. In the *Preliminary Discourse*, Herschel maintained that "the Divine Author of the universe cannot be supposed to have laid down particular laws, enumerating all individual contingencies, which his materials have understood and obey . . . but rather, by creating them, endued with certain fixed qualities and powers, he has impressed them in their origin with the spirit, not the letter, of his law" (37).

92. Darwin, *Autobiography*, 93–94.

93. Michael Ruse, *Darwin and Design: Does Evolution Have a Purpose?* (Cambridge, MA: Harvard University Press, 2003), 112.

94. Elliott Sober, "Darwin and Naturalism," in Sober, *Did Darwin Write the Origin Backwards? Philosophical Essays on Darwin's Theory* (Amherst, NY: Prometheus Books, 2011), 121–52, quotation from 128.

95. See Sydney Ross, "Scientist: The Story of a Word," *Annals of Science* 18, no. 2 (1964): 65–85.

96. Whewell, *History of the Inductive Sciences*, 1:44.

97. Ross, "Scientist: The Story of a Word," 72.

98. Darwin, *Origin of Species*, 62. The emphasis is mine.

99. Charles Darwin, *Charles Darwin's Natural Selection, Being the Second Part of His Big Species Book Written from 1856 to 1858*, ed. R. C. Stauffer (Cambridge: Cambridge University Press, 1975), 224.

100. Lyell mentions Brocchi's theory in *Principles of Geology*, 2:128. Darwin's entry marks the first time he speculates openly about species change, sometime in March 1837; see Darwin, *Red Notebook* (MS, 129), in *Charles Darwin's Notebooks*, 62. The final quotation is from *Journal of Researches*, 212.

101. Darwin, *Notebook B* (MS, 38–39), in *Charles Darwin's Notebooks*, 180.

102. Darwin, *Notebook C* (MS, 63), in *Charles Darwin's Notebooks*, 259.

103. Ruse, *The Darwinian Revolution*, 175–76. Ruse contends that given the background of reading Whewell and Herschel, and Malthus's quantitative analysis, "Darwin was becoming aware of something that would be a prime candidate for a scientific evolutionary mechanism" (176).

104. Lyell, *Principles of Geology*, 2:41.

105. Darwin's notebooks indicate he read carefully the works of Sebright and Wilkin-

son, especially, John Sebright, *The Art of Improving the Breeds of Domestic Animals* (London: Howlett and Brimmer, 1809); and John Wilkinson, *Remarks on the Improvement of Cattle, etc. in a Letter to Sir John Sanders Sebright, Bart. M.P.* (Nottingham, 1820).

106. Ruse nicely explains how the breeders contributed to Darwin's understanding of the nature of artificial selection. See Michael Ruse, "Charles Darwin and Artificial Selection," *Journal of the History of Ideas* 36 (1975): 339–50.

107. Darwin, *Notebook D* (MS, 20), in *Charles Darwin's Notebooks*, 337.

108. Lyell, *Principles of Geology*, 2:131.

109. Darwin, *Notebook C* (MS, 73), in *Charles Darwin's Notebooks*, 262.

110. Darwin, *Notebook D* (MS, 135e), in *Charles Darwin's Notebooks*, 375–76.

111. Darwin, *Notebook N* (MS, 1–3), in *Charles Darwin's Notebooks*, 563–64.

112. Charles Darwin, *Charles Darwin's Natural Selection*, 468.

113. William Kirby and William Spence, *Introduction to Entomology*, 2nd ed., 4 vols. (London: Longman, Hurst, Rees, Orne, and Brown, 1818), 2:471.

114. I discuss Darwin's work on the social instincts and the application he made to human beings in *Darwin and the Emergence of Evolutionary Theories of Mind and Behavior* (Chicago: University of Chicago Press, 1987), 139–52 and 212–19.

115. Darwin, *Notebook B* (MS, 5), in *Charles Darwin's Notebooks*, 171.

116. Darwin, *Notebook E* (MS, 48–49), in *Charles Darwin's Notebooks*, 409. Double wedge quotes indicate a passage later added.

117. Ibid.

118. Darwin, *Origin of Species*, 490.

119. Darwin first introduced the model in the "Essay of 1842," in *Foundations of the Origin of Species*, 6, and elaborated it in the above passage from the "Essay of 1844," in ibid., 85.

120. Darwin, "Essay of 1842," in ibid., 6.

121. See, Darwin, "Essay of 1844," in ibid., 85; *Big Species Book*, 224–25; *Origin of Species*, 83–84.

122. Ibid.

123. Darwin, *Big Species Book*, 225. There is an assumption that Darwin abandoned the notion that natural selection could produce "perfect adaptations" by the 1850s, at least this is Ospovat's view and it has been adopted by most other scholars. See Dov Ospovat, *The Development of Darwin's Theory* (Cambridge: Cambridge University Press, 1981), 4. As the citation from *Big Species Book* shows, this is certainly not the case. Moreover, the notion of "infinitely better adapted" (asserted in the quotation above) is the functional equivalent of perfect. I wouldn't deny Darwin sometimes expresses the idea that there are degrees of perfection, but he seems to have held that notion from the very beginning.

124. Darwin, *Origin of Species*, 83, 84, 149, 194, 201, and 489.

125. Ibid., 489.

126. Darwin, *Origin*, 80.

127. Thomas Henry Huxley, "Darwin on the Origin of Species," *Westminster Review*, n.s., 17 (1860): 541–70; quotation from 569.

128. [Fleeming Jenkin], "The Origin of Species," *North British Review* 46 (1867): 277–318.

129. Darwin, *Origin: Variorum Text*, 179.

130. The first two chapters of Darwin's manuscript are missing, but presumably dealt with domestic selection. Chapter 3 is on variation in nature (chapter 2 of the *Origin*) and Chapter 4 on crossing of organic beings. Chapter 4 was dropped from the *Origin* and much of it appeared in *Animals and Plants under Domestication*.

131. Alfred Russel Wallace to Darwin (July 2, 1866), in *Correspondence of Charles Darwin* 14:227–29.

132. Darwin, *Origin of Species*, 336–37.

133. Ibid., 489.

134. See Michael Ruse, "Darwin and Group Selection," *Annals of Science* 37 (1980): 615–30. For the most recent assertion of this position, see Michael Ruse, "Charles Robert Darwin and Alfred Russel Wallace: Their Dispute over the Units of Selection," *Theory in Biosciences* 132 (2013): 215–24.

135. See W. D. Hamilton, "The Genetical Evolution of Social Behavior," *Journal of Theoretical Biology* 7 (1964): 1–16, 17–51.

136. Ibid., 87.

137. Darwin, *Origin: Variorum Text*, 172.

138. Alfred Russel Wallace, "The Origin of Human Races and the Antiquity of Man Deduced from the Theory of 'Natural Selection,'" *Anthropological Review* 2 (1864): clviii–clxxxvii.

139. Charles Darwin to Alfred Wallace (May 28, 1864), in *Correspondence of Charles Darwin*, 12:216–17.

140. Wallace, "The Origin of Human Races," clxii.

141. Darwin, *Descent of Man*, 1:166.

142. Ruse makes this point in several places, including "Charles Darwin and Alfred Russel Wallace," 221: "Wallace, the man who has now [1858] has been a committed socialist, makes it clear that he is thinking in terms of groups rather than individuals." I do not think Ruse is correct about this, though it has become a standard way of regarding Wallace's views in the 1858 joint paper. I will take this up in my response to Ruse.

143. Darwin, *Origin*, 488.

144. Adam Sedgwick, "Objections to Mr. Darwin's Theory of the Origin of Species," in *Darwin and His Critics*, ed. David Hull (Cambridge, MA: Harvard University Press, 1973), 159–66; quotation from 165. Hull's volume includes a number of reviews of the *Origin* with his commentary.

145. Ibid., 165.
146. Darwin to Alfred Russel Wallace (March 12–17, 1867), in *Correspondence of Charles Darwin*, 15:141.
147. Alfred Wallace to Charles Lyell (April 28, 1869), in the Lyell Correspondence, BD 25, I., American Philosophical Society, Philadelphia.
148. [St. George Jackson Mivart], *Quarterly Review* 131 (1871): 47–90; reference on 72. Mivart ended his review (90) with another sneer: "We sincerely trust Mr. Darwin may yet live to furnish us with another work, which, while enriching physical science, shall not, with needless opposition, set at naught the first principles of both philosophy and religion."
149. Charles Darwin, *The Descent of Man and Selection in Relation to Sex*, ed. James Moore and Adrian Desmond, 2nd ed. (London: Penguin, [1879] 2004), 91n.
150. "Mr. Darwin on the Descent of Man," *Times* (London) (April 7, 1871).
151. "Darwin on the Descent of Man," *Edinburgh Review* 134 (1871): 195–235; quotation from 195–96.
152. Darwin, *Descent of Man*, 1:149–50.
153. Ibid., 2:328.
154. Ibid., 2:390.
155. Ibid., 1:161.
156. Ibid., 1:155.
157. Ibid., 157.
158. Ibid., 1:70–71.
159. Kant's book was translated as Immanuel Kant, *Metaphysics of Ethics*, trans. J. W. Semple (Edinburgh: T. Clark, 1836).
160. Cobbe later emphasized the difference between the approach of Darwin and Kant in her critique of Darwin's theory of morality. See Frances Power Cobbe, "Darwinism in Morals," *Theological Review* 8 (1871): 167–92. The first part of the article endorses Darwin's general theory, finding it perfectly compatible with the idea that God orchestrated the gradual evolution of human beings from lower forms. But she objected to Darwin's theory of morals, since it did not seem sufficient to account for feelings of conscious regret for transgressions and the universal character of right action.
161. Darwin, *Descent of Man*, 1:71–73.
162. Ibid., 1:72.
163. Ibid.
164. Ibid., 1:166 and a similar passage on 162.
165. Ibid., 1:155.
166. Darwin, *Origin of Species: Variorum Text*, 172.
167. Darwin, *Descent of Man*, 1:163.
168. Ibid., 1:164.
169. Ibid., 1:98.
170. He was probably also influenced quite early on by his reading of James Mack-

intosh's *Dissertation on the Progress of Ethical Philosophy*, which drew upon another strain of British tradition, that of the moral sense school of Shaftesbury and Hutchinson. Yet, he still attempted to work out his moral theory in the late 1830s and early 1840s within the Paleyesque conception of utility. See above and my *Darwin and the Emergence of Evolutionary Theories of Mind and Behavior*, 114–19.

171. I have discussed Mill's effort in *Darwin and the Emergence of Evolutionary Theories of Mind and Behavior*, 233–42.

172. See Darwin's *Old and Useless Notes* (MS, 33), in *Charles Darwin's Notebooks*, 610. See also, John Stuart Mill, "Review of the Works of Samuel Taylor Coleridge," *London and Westminster Review* 33 (1840): 257–302.

173. Darwin, *Descent of Man*, 1:97–98.

174. Ruse, "Darwin and Group Selection." See my discussion above. The evidence is clear that Darwin was indeed a group selectionist.

175. Michael Ruse and E. O. Wilson, "Moral Philosophy as Applied Science," *Philosophy* 61 (1986): 173–92; quotation from 179.

176. Ibid., 186.

177. Ibid.

RESPONSE TO RUSE

1. Charles Darwin, *Autobiography*, ed. Nora Barlow (London: Collins, 1958), 119.

2. Edward Hallett Carr, *What Is History?* (New York: Vintage, 1961), 24–26.

3. Ruse, 2. In subsequent notes, when referring to Ruse's essay in this volume, I will simply use his name and a page number.

4. Ibid.

5. Darwin, *Origin of Species*, 62.

6. Ibid., 67.

7. Ibid., 62.

8. Ruse, 46.

9. Darwin to Joseph Hooker (June 8, 1858), in *Correspondence of Charles Darwin*, 7:102.

10. Darwin, *Autobiography*, 120–21.

11. Darwin, *Big Species Book*, 238.

12. Darwin, *Origin of Species*, 112. I have discussed Darwin's principle of divergence at length in "Darwin's Principle of Divergence: Why Fodor Was Almost Right," in my *Was Hitler a Darwinian*, 55–89.

13. Ruse, 49.

14. This is an example that Henri Milne-Edwards mentioned in his *Introduction à la zoologie générale* (Paris: Victor Masson, 1851), 43–44. Darwin used this same example in his *Big Species Book*, 233. He changed the example in the *Origin* (116),

where he specified that the division of labor was between two different organisms, one with a stomach designed for digesting vegetation, the other with one for meat. The change indicates he sensed a problem with the application.

15. Darwin wrote to his friend Joseph Hooker to say that he thought Milne-Edwards's notion of the division of labor was the surest criterion of highness in the scale of life. See Darwin to Joseph Hooker (June 27, 1854), in *Correspondence of Charles Darwin*, 5:197.

16. This resolution to the problem of the divergence of organisms is one for which Darwin himself argued in his 1844 essay (*Foundations of the Origin of Species*, 213). He later thought that he had missed something, and then proposed the notion of selecting for extremes as the added feature. I have explored the construction of Darwin's principle of divergence in "Darwin's Principle of Divergence: Why Fodor Was Almost Right," in *Was Hitler a Darwinian*, 55–89.

17. Thomas Henry Huxley, "The Genealogy of Animals," in Thomas Henry Huxley, *Collected Essays*, 8 vols. (London: McMillan & Co., 1893), 2:110.

18. Ibid.

19. Immanuel Kant, *Kritik der Urteilskraft*, A347, B351.

20. See William Paley, *Natural Theology; or, Evidences of the Existence and Attributes of the Deity*, 12th ed. (London: J. Faulder [1802], 1809), 40–42.

21. To remind the reader, Darwin, in his *Autobiography* (92–93), said that up through the time of the publication of the *Origin of Species*, he believed the universe governed by an intelligence comparable to our own minds.

22. Darwin, *Notebook E* (MS, 48–49), in *Charles Darwin's Notebooks*, 409. See also my essay.

23. Michael Ruse, *The Darwinian Revolution* (Chicago: University of Chicago Press, 1979), 199.

24. Stephen Jay Gould, "Eternal Metaphors of Palaeontology," in *Patterns of Evolution as Illustrated in the Fossil Record*, ed. A. Hallan (New York: Elsevier, 1977), 13.

25. Peter Bowler, *Theories of Human Evolution* (Baltimore, MD: Johns Hopkins University Press, 1986), 41.

26. Darwin, *Origin of Species*, 490.

27. See my exchanges of letters with Richard Lewontin who made the claim that the term "evolution" was absent from Darwin's corpus: "Darwin and Progress," *New York Review of Books* (December 15, 2005).

28. Michael Ruse, *Monad to Man: The Concept of Progress in Evolutionary Biology* (Cambridge, MA: Harvard University Press, 1996), 166.

29. Ruse, 54.

30. Ibid., 55.

31. Charles Darwin, *The Origin of Species by Charles Darwin: A Variorum Text*, ed. Morse Peckham (Philadelphia: University of Pennsylvania Press, 1959), 221.

32. Darwin, *Notebook E* (MS, 48–49), in *Charles Darwin's Notebooks*, 409.

33. Darwin, DAR 205.9 (2), MS, 200, held by the Manuscript Room, Cambridge University Library.

34. Charles Darwin, *On the Origin of Species*, 3rd ed. (London: Murray, 1861), 133. He further added (134): "If we look at the differentiation and specialisation of the several organs of each being, when adult (and this will include the advancement of the brain for intellectual purposes), as the best standard of highness of organisation, natural selection will clearly lead towards highness."

35. Darwin, *Notebook B* (MS, 49), in the *Notebooks of Charles Darwin*, 182.

36. Darwin, *Origin of Species*, 61.

37. I discuss the case of the social insects in more detail below.

38. Ruse, 63.

39. Darwin, *Descent of Man*, 1:316.

40. Charles Darwin, *The Descent of Man and Selection in Relation to Sex*, 2nd ed., 2 vols. (London: Murray, 1874), 259–60; my emphasis.

41. Wallace was later perfectly sanguine about using artificial selection as a model for natural selection. See his *Darwinism: An Exposition of the Theory of Natural Selection, with Some of Its Applications* (Cambridge: Cambridge University Press, 1889), 83–101.

42. See, for example, Peter Bowler's discussion in his *Evolution, the History of an Idea*, 3rd ed. (Berkeley: University of California Press, 2003), 173–76. "It can be argued that Wallace was not thinking of the key Darwinian mechanism—the struggle for existence between individuals within a population—but focused instead on competition between distinct varieties or subspecies" (175).

43. Wallace quoted by Ruse in "Charles Robert Darwin and Alfred Russel Wallace: Their Dispute over the Units of Selection," *Theory in Biosciences* 132 (December 2013): 215–24. The original passage is from Wallace's essay originally read to the Linnaean Society: "On the Tendency of Varieties to Depart Indefinitely from the Original Type," in Charles Darwin and Alfred Wallace, "On the Tendency of Species to form Varieties; and on the Perpetuation of Varieties and Species by Natural means of Selection," *Journal of the Proceedings of the Linnaean Society of London. Zoology* 3 (1858): 45–62; quotation from 58.

44. Darwin, *Origin of Species*, 52.

45. Wallace, *Darwinism*, 106.

46. Alfred Russel Wallace to Darwin (March 1, 1868), in *Correspondence of Charles Darwin*, 219–20.

47. Ibid., 220–21.

48. Wallace, *Darwinism*, 152–79. Wallace provided many examples, drawn from Darwin's work, that gave added weight to his friend's contingency explanation. However, Wallace still attempted to imagine how natural selection could aid the establishment of hybrid sterility. After going on for some pages, the com-

plexity and ad hoc assumptions of his argument drove him to throw up his hands and say that a simpler argument could be found in the brief exposition he had originally given to Darwin in a letter some twenty years prior; he then produced that exposition in a footnote (179). The shortened version was not any clearer.

49. William Paley, *Moral and Political Philosophy*, in *The Works of William Paley*, new ed., 7 vols. (London: C. and J. Rivington, 1825), 4:54.

50. Darwin's manuscript on Mackintosh is included in the bundle of notes he labeled *Old and Useless Notes* (MS, 42–55), in *Charles Darwin's Notebooks*, 618–29. Darwin epitomizes his own position in the quotation from Mackintosh: "The pleasure which results when the object is attained (the gratification of one's offspring) is not the aim of the agent, for it does not enter into his contemplation" (627).

51. Ibid., 1:155.

52. Alfred Russel Wallace, "The Origin of Human Races and the Antiquity of Man Deduced from the Theory of 'Natural Selection,'" *Journal of the Anthropological Society of London* 2 (1864): clviii–clxx; quotation from clxviii.

53. Darwin, *Descent of Man*, 1:103.

54. *Journal of Researches into the Natural History and Geology of the Countries Visited during the Voyage of H.M.S. Beagle Round the World, under the Command of Capt. Fitz Roy, R.N.*, 2nd ed. (London: John Murray, 1845), 499–500.

55. Darwin, *Descent of Man*, 1:98.

56. Ibid., 155.

REPLY TO RICHARDS

1. Letter from Karl Marx to Friedrich Engels, June 18, 1862.

2. Robert J. Richards, *Was Hitler a Darwinian? Disputed Questions in the History of Evolutionary Theory* (Chicago: University of Chicago Press, 2013), 92.

3. Charles Darwin, *The Autobiography of Charles Darwin, 1809–1882: With the Original Omissions Restored; Edited and with Appendix and Notes by His Grand-Daughter Nora Barlow* (London: Collins, 1958).

4. Paul H. Barrett, Peter J. Gautrey, Sandra Herbert, David Kohn, and Sydney Smith, eds., *Charles Darwin's Notebooks, 1836–1844* (Ithaca, NY: Cornell University Press, 1987), M: 27, Darwin's spelling.

5. A. Rupert Hall, *The Revolution in Science, 1500–1750* (London: Longman, 1983).

6. From "Tintern Abbey," by William Wordsworth, "Composed a few miles above Tintern Abbey, on revisiting the banks of the Wye during a tour. July 13, 1798."

7. Mention has already been made of Elliott Sober, *Did Darwin Write the "Origin" Backwards* (Buffalo, NY: Prometheus Books, 2010).

8. Good scientific overviews are Stuart A. West, A. S. Griffin, and Andy Gardner,

"Social Semantics: How Useful Has Group Selection Been?," *Journal of Evolutionary Biology* 21 (2007): 374–85, and Stuart A. West and Andy Gardner, "Adaptation and Inclusive Fitness," *Current Biology* 23 (2013): R577–R584.

9. Sam Schweber, "Darwin and the Political Economists: Divergence of Character," *Journal of the History of Biology* 13 (1980): 195–289; "The Wider British Context in Darwin's Theorizing," in *The Darwinian Heritage*, ed. David Kohn (Princeton, NJ: Princeton University Press, 2005), 35–69.

10. J. R. M'Culloch, *The Principles of Political Economy* (London: Alex, Murray and Son, 1825), 36.

11. Michael Ruse, "Charles Darwin and Group Selection," *Annals of Science* 37 (1980): 615–30.

12. Letter from Darwin to Joseph D. Hooker, December 14, 1859, in Charles Darwin, *The Collected Correspondence of Charles Darwin* (Cambridge: Cambridge University Press, 1985–), 7:431.

13. Darwin, *Autobiography*, 125.

14. Very helpful is Robert J. Richards, *The Meaning of Evolution: The Morphological Construction and Ideological Reconstruction of Darwin's Theory* (Chicago: University of Chicago Press, 1992); see also *The Romantic Conception of Life: Science and Philosophy in the Age of Goethe* (Chicago: University of Chicago Press, 2003), and *The Tragic Sense of Life: Ernst Haeckel and the Struggle over Evolutionary Thought* (Chicago: University of Chicago Press, 2008).

15. I quote here from Richards's main essay in this volume.

16. Richards, quoting Barrett et al., *Notebooks*, E: 49.

17. Charles Darwin, *On the Origin of Species* (London: John Murray, 1859), 440.

18. Edwin S. Russell, *Form and Function: A Contribution to the History of Animal Morphology* (London: John Murray, 1916), 126, quoting Karl Ernst von Baer.

19. Barrett et al., *Notebooks*, E: 49.

20. Darwin, *Autobiography*, 85, 138.

21. Barrett et al., *Notebooks*, M: 40.

22. William Wordsworth, preface to the *Lyrical Ballads*, quoted by Gillian Beer, "Darwin's Reading and the Fictions of Development," in *The Darwinian Heritage*, ed. David Kohn (Princeton, NJ: Princeton University Press, 1985), 546.

23. Darwin, *Autobiography*, 104.

24. Barrett et al., *Notebooks*, B: 19.

25. Ibid., B: 161.

26. Ibid., B: 163.

27. Ibid., C: 48e.

28. Ibid., D: 115.

29. Thomas Henry Huxley, "On the Theory of the Vertebrate Skull: Croonian Lecture Delivered before the Royal Society, June 17, 1858," *Proceedings of the Royal Society* 9 (1857–59): 381–457.

30. Malcolm Nicholson, "Historical Introduction," to Alexander von Humboldt, *Personal Narrative of a Journey to the Equinoctial Regions of the New Continent*, trans. Jason Wilson (London: Penguin, 1995), ix.

31. Ibid., xx.

32. Darwin, *Autobiography*, 68.

33. Charles Darwin, *Journal of Researches into the Natural History and Geology of the Countries Visited during the Voyage of H.M.S. Beagle round the World*, 2nd ed. (London: John Murray, 1845), 503.

34. Ibid., 11.

35. Ibid., 503.

36. Caroline Darwin to Darwin, October 28, 1833, in Darwin, *Correspondence*, 1:224.

37. David R. Stoddart, ed., "Coral Islands by Charles Darwin," *Atoll Research Bulletin* 88 (December 1962): 1–20, 17. This essay is dated 1835, and has the Cambridge Darwin Collection catalog number CUL-DAR 41:1–12.

38. Ibid. In the original essay the quotation is in the original French. There is a one-page translation in the Darwin Collection, CUL-DAR 42:23.

39. Letter from Darwin to J. D. Hooker, August 12, 1881, *Darwin Correspondence Project*, letter 13288. Reprinted in Francis Darwin and A. C. Steward, *More Letters of Charles Darwin* (London: Murray, 1903), 2:26.

40. Letter from Darwin to Charles Lyell, October 8, 1845, in Darwin, *Correspondence*, 3:258.

41. Letter from Darwin to J. D. Hooker, August 6, 1881, *Darwin Correspondence Project*, letter 13277. Also in Francis Darwin, *Life and Letters of Charles Darwin* (London: John Murray, 1887), 3:247.

42. Alexander von Humboldt and Aimé Bonpland, *Personal Narrative of Travels to the Equinoctial Regions of the New Continent, during the Years 1799–1804*, trans. Helen Maria Williams, 7 vols. (London: Longman, Hurst, Rees, Orme & Brown, 1819–29), 4:217.

43. Barrett et al., *Notebooks*, M: 38–39.

44. Erasmus Darwin, *Temple of Nature* (London: J. Johnson, 1803), footnote to line 176.

45. Ibid., footnote to line 207.

46. Humboldt and Bonpland, *Narrative*, 3:491.

47. Ibid., 5:565.

48. Darwin marginalia, to Humboldt and Bonpland, *Narrative*, 5:590.

49. Ibid., end of volume 6.

50. Darwin, *Origin*, 84.

51. F. Van Dyke, D. C. Mahan, J. K. Sheldon, and R. H. Brand, *Redeeming Creation: The Biblical Basis for Environmental Stewardship* (Downer's Grove, IL: InterVarsity Press, 1996), 53.

52. Darwin and Wallace, *Evolution*, 45–46; this is from the "Sketch."

53. Ibid., 154; this is from the "Essay."

54. Letter from Darwin to Charles Lyell, February 15, 1860, in Darwin, *Correspondence*, 8:87.

55. Eduard J. Dijksterhuis, *The Mechanization of the World Picture* (Oxford: Oxford University Press, 1961).

56. Michael Ruse, *Science and Spirituality: Making Room for Faith in the Age of Science* (Cambridge: Cambridge University Press, 2010).

57. Darwin, *Voyage*, 503.

58. John Chancellor, "Humboldt's Personal Narrative and Its Influence on Darwin," http://darwin-online.org.uk/EditorialIntroductions/Chancellor_Humboldt .html.

59. Humboldt and Bonpland, *Narrative*, 4:505–6.

60. Darwin, *Autobiography*, 85.

61. Letter from Darwin to John S. Henslow, November 24, 1832, in Darwin, *Correspondence*, 1:279. The Milton passage is: "Him there they found / Squat like a toad, close at the ear of Eve, / Assaying by his devilish art to reach / The organs of her fancy, and with them forge / Illusions, as he list, phantasms and dreams." It is from *Paradise Lost*, 4:799–803.

62. *Paradise Lost*, 7:278–90.

63. Ibid., 7:391–98.

64. Ibid., 7:519–31.

65. Actually, Milton himself denied the Trinity.

66. Leave Darwin slumbering, but realize that the clash between Richards and myself is no mere historical exercise. Many of the things over which we argue are still live issues in evolutionary circles today. A recent multiauthored article in the leading journal *Nature* shows that there is still strong disagreement over such things as the nature and relevance of development in the evolutionary story. Our interest is in Darwin and our exchange is about him; but we are sensitive to the ongoing implications of our thinking. See Kevin Laland, Tobias Uller, Marc Feldman, Kim Sterelny, Gerd B. Müller, Armin Moczek, Eva Jablonka, et al. "Does Evolutionary Theory Need a Rethink?," *Nature* 514 (2013): 162–64.

EPILOGUE

1. Alfred North Whitehead, *Process and Reality* (New York: Free Press, 1979), 39.

2. For a brief but authoritative history of post-Mendelian genetics, see Elof Carlson, *Mendel's Legacy: The Origin of Classical Genetics* (Cold Spring Harbor, NY: Cold Spring Harbor Laboratory Press, 2004).

3. Erik Nordenskiöld, *History of Biology* (New York: Knopf, 1935 [1920–24]), 477: "Modern critics have often asked themselves how it is that a hypothesis like Darwin's, based on such weak foundations, could all at once win over to its

side the greater part of contemporary scientific opinion. . . . From the beginning Darwin's theory was an obvious ally to liberalism; it was at once a means of elevating the doctrine of free competition, which had been one of the most vital corner-stones of the movement of progress, to the rank of natural law, and similarly the leading principle of liberalism, progress, was confirmed by the new theory."

4. Michael Ruse, "Population Genetics," in *The Cambridge Encyclopedia of Darwin and Evolutionary Thought*, ed. Michael Ruse (Cambridge: Cambridge University Press, 2013), 273–81.

5. J. D. Watson and F. H. Crick, "Molecular Structure of Nucleic Acids: A Structure for Deoxyribose Nucleic Acid," *Nature* 171 (1953): 737–38.

6. August Weismann, "The Supposed Transmission of Mutilations," in his *Essays upon Heredity and Kindred Biological Problems*, trans. and ed. E. Poulton et al., 2 vols. (Oxford: Clarendon Press, 1889), 1:421–48 (especially 431–33).

7. See August Weismann, *The Germ-Plasm: A Theory of Heredity*, trans. W. N. Parker and H. Rönnfeldt (New York: Scribner's Son, 1893).

8. Carlson, *Mendel's Legacy*, 295–97.

9. William Hamilton, "The Genetical Evolution of Social Behavior, I and II," *Journal of Theoretical Biology* 7 (1964): 1–16; 17–52.

10. See Richards, *Darwin and the Emergence of Evolutionary Theories of Mind and Behavior*, 541.

11. Robert Trivers, "The Evolution of Reciprocal Altruism," *Quarterly Review of Biology* 46 (1971): 35–57.

12. E. O. Wilson, *Sociobiology: The New Synthesis* (Cambridge, MA: Harvard University Press, 1975); Richard Dawkins, *The Selfish Gene* (Oxford: Oxford University Press, 1976).

13. George C. Williams, *Adaptation and Natural Selection* (Princeton, NJ: Princeton University Press, 1966). For a searching examination of Wynn Edwards's work, see Mark Borrello, *Evolutionary Restraints: The Contentious History of Group Selection* (Chicago: University of Chicago Press, 2010).

14. See, for instance, Michael J. Wade, "A Critical Review of the Models of Group Selection," *Quarterly Review of Biology* 53 (1978): 101–14; David Sloan Wilson and E. O. Wilson, "Rethinking the Theoretical Foundation of Sociobiology," *Quarterly Review of Biology* 82 (2007): 327–48; and Martin A. Nowak, Corina E. Tarnita, and Edward O. Wilson, "The Evolution of Eusociality," *Nature* 466 (2010): 1057–62. Many, probably most, evolutionary biologists differ very strongly with these claims. See Patrick Allen et al., "Inclusive Fitness and Eusociality," *Nature* 471 (2011): E1–E4. There were 150 signatories to this letter, which was one of five in the same issue.

15. In the first edition of the *Origin of Species* (285–87), Darwin calculated that the erosion of certain cliffs along the English coast would have taken at least 300

million years, and this was only a more obvious indicator of the great age of the earth. But when Kelvin put the upper limit on the earth's age at about 200 million years, Darwin began recalculating. Already in the second edition (1860), he reduced the time of cliff erosion to 150 to 100 million years; and by the fifth edition (1869), he suggested that the geologist Croll might be right that the initial layers of the Cambrian formation, when life first appeared, occurred only 60 million years ago. In the first edition, Darwin urged that the imagination could not comprehend the length of 100 million years (481); by the fifth edition, the imagination could not comprehend the span of 1 million years and what measure of evolution could have occurred in that time. See Charles Darwin, *On the Origin of Species: A Variorum Edition*, ed. Morris Peckham (Philadelphia: University of Pennsylvania Press, 1959), 484–86.

16. See Robert J. Richards, *The Meaning of Evolution: The Morphological Construction and the Ideological Reconstruction of Darwin's Theory* (Chicago: University of Chicago Press, 1992, 136–43. Note that Ruse and Richards differ sharply over the precise meaning for Darwin of the "principle of recapitulation."

17. For studies of the recent developments in paleontology, see David Sepkoski and Michael Ruse, eds., *The Paleobiological Revolution: Essays on the Growth of Modern Paleontology* (Chicago: University of Chicago Press, 2009); David Sepkoski, *Rereading the Fossil Record: The Growth of Paleobiology as an Evolutionary Discipline* (Chicago: University of Chicago Press, 2012).

18. Charles Darwin, *On the Various Contrivances by Which British and Foreign Orchids Are Fertilised by Insects, and On the Good Effects of Intercrossing* (London: Murray, 1862), 197–98: "In several flowers sent me by Mr. Bateman I found the nectaries eleven and a half inches long, . . . In Madagascar there must be moths with prosbosces capable of extension to a length of between ten and eleven inches!"

19. Ibid.

20. See V. de Buffrénil, J. O. Farlow, and A. de Ricqlès, "Growth and Function of *Stegosaurus* Plates: Evidence from Bone Histology," *Paleobiology* 12 (1986): 459–73.

21. Darwin, *Origin of Species*, 450.

22. The history of evolutionary developmental biology and its startling findings are clearly depicted in Sean B. Carroll, *Endless Forms Most Beautiful: The New Science of Evo Devo* (New York: Norton, 2005).

23. Elizabeth Pennisi, "Genomics: Encode Project Writes Eulogy for Junk DNA," *Science* 337 (September 7, 2012): 1159–61.

24. The type specimen was discovered in the Neanderthal valley in 1856. Earlier specimens were found in what is now Belgium (1829) and Gibraltar (1848).

25. Robert J. Richards, *The Tragic Sense of Life: Ernst Haeckel and the Struggle over Evolutionary Thought* (Chicago: University of Chicago Press, 2008), 252–55;

Michael Ruse, *The Philosophy of Human Evolution* (Cambridge: Cambridge University Press, 2012), 48.

26. See Richard Leakey and Roger Lewin, *Origins: The Emergence and Evolution of Our Species and Its Possible Future* (New York: Penguin 1991); *Origins Reconsidered: In Search of What Makes Us Human* (New York: Anchor Books, 1993); and Donald Johanson and Maitland Edey, *Lucy: The Beginnings of Humankind* (New York: Simon & Schuster, 1990).

27. Richard Green et al., "A Draft Sequence of the Neanderthal Genome," *Science* 328 (May 7, 2010): 710–20.

28. Ewen Callaway, "The Discovery of Homo Floresiensis: The Tales of the Hobbit," *Nature* 514 (October 23, 2014): 422–26.

29. See, for example, John Tooby and Leda Cosmides, "The Psychological Foundations of Culture," in *The Adapted Mind*, eds. Jerome Barkow, Leda Cosmides, and John Tooby (New York: Oxford University Press, 1992), 19–136.

30. Frans de Waal has performed many experiments on apes and monkeys that display the rudiments of moral concern. See, for example, his *The Age of Empathy: Nature's Lessons for a Kinder Society* (New York: Crown, 2009).

31. Marc Hauser, Liane Young, and Fiery Cushman, "Reviving Rawls' Linguistic Analogy: Operative Principles and the Causal Structure of Moral Actions," in *Moral Psychology*, ed. W. Sinnott-Armstrong, 3 vols. (Cambridge: MIT Press, 2007), 2:107–44.

32. Theodosius Dobzhansky, "Nothing in Biology Makes Sense Except in the Light of Evolution," *American Biology Teacher* 35 (1973): 125–29.

33. C. D. Broad, *The Mind and Its Place in Nature* (London: Kegan Paul, 1925).

34. Andreas Vesalius, *De Corporis Humani Fabrica, Libri Septem* (Basel: Oporinus, 1543), 164.

35. Richards, *The Tragic Sense of Life*, 368–71.

36. Daniel Dennett, *Consciousness Explained* (New York: Little Brown, 1991).

37. See Thomas Henry Huxley, "On the Hypothesis That Animals Are Automata and Its History," *Fortnightly Review* 22 (1874): 555–89.

38. At least they endorse the argument in principle. Popper quotes Donald Campbell, who exploits the James argument ("'Downward Causation' in Hierarchically Organized Biological Systems," in *Studies in the Philosophy of Biology*, ed. Francisco Ayala and Theodosius Dobzhansky [London: Macmillan, 1974], 179–86). See Karl Popper and John Eccles, *The Self and Its Brain* (New York: Springer-Verlag, 1977). Their book is a sustained argument for the emergence of consciousness in organic life: "The most reasonable view seems to be that consciousness is an emergent property of animals arising under the pressure of natural selection" (29).

39. See Robert J. Richards, *Darwin and the Emergence of Evolutionary Theories of Mind and Behavior* (Chicago: University of Chicago Press, 1987), 430–35.

40. See William James, "Does Consciousness Exist?," in *The Works of William James: Essays in Radical Empiricism*, ed. Frederick Burkhardt (Cambridge, MA: Harvard University Press, 1976), 3–20; and Bertrand Russell, *The Analysis of Mind* (London: George Allen & Unwin, 1978 [1921]).

41. William James, *The Principles of Psychology*, 2 vols. (New York: Henry Holt, 1890), 1:145–62.

42. James, "Does Consciousness Exist?," 19.

43. This conclusion forms the subtitle of Nagel's recent book: *Mind and Cosmos: Why the Materialist Neo-Darwinian Conception of Nature Is Almost Certainly False* (Oxford: Oxford University Press, 2012).

44. Ibid., 27.

45. Ibid. 41.

46. Colin McGinn argues the "new mysterian" line, namely, that although consciousness obviously exists, we may never solve the mystery of its full nature and connection to the material world. See Colin McGinn, *The Mysterious Flame: Conscious Minds in a Material World* (New York: Basic Books, 2000). While the evolutionary theorist can go some distance in explaining the nature of consciousness, there are nonetheless certain brute facts of the matter—for example, that neural cells produce consciousness.

47. Nagel, *Mind and Cosmos*, 46.

48. Ibid., 44, 45, 50, 52, 60.

49. Ibid., 48.

50. Ibid., 83.

51. Michael Ruse traces the history of the design argument and shows the ways in which Darwin's theory instantiates the idea of design without a designing mind. See his *Darwin and Design: Does Evolution Have a Purpose?* (Cambridge, MA: Harvard University Press, 2003), especially 111–28.

52. Darwin, *Autobiography*, 92–93.

53. Darwin to John Fordyce (May 7, 1879), in *Life and Letters of Charles Darwin*, 1:304.

54. Darwin, *Descent of Man*, 1:65–69.

55. E. O. Wilson, *Sociobiology* (Cambridge, MA: Harvard University Press, 1975), 559–62. See also David Sloan Wilson, *Darwin's Cathedral* (Chicago: University of Chicago Press, 2002).

56. James Leuba, *The Belief in God and Immortality: A Psychological, Anthropological and Statistical Study* (Boston: Sherman, French, & Co., 1916), 253–55 especially.

57. See Edward Larson and Larry Witham, "Scientists Are Still Keeping the Faith," *Nature* 386 (April 3, 1997): 435–36; and "Leading Scientists Still Reject God," *Nature* 394 (July 23, 1998): 313. A more recent study indicates that elite scientists fell into the category of atheist or agnostic to the tune of 64 percent of such scientists, while only 6 percent of the American public did. Nine percent of elite

scientists professed a belief in God, while 63 percent of the American public expressed belief in a personal God. See Elaine Howard Ecklund, *Science vs. Religion: What Scientists Really Think* (Oxford: Oxford University Press, 2010), 16.

58. See Robert L. Carroll, *Vertebrate Paleontology and Evolution* (New York: Freeman and Co., 1988), 361–400.

59. Michael H. Day gives a complete description of hominin discoveries up to the date of his *Guide to Fossil Man*, 4th ed. (Chicago: University of Chicago Press, 1986). See also Ian Tattersall, *The Fossil Trail: How We Know What We Think We Know about Human Evolution* (Oxford: Oxford University Press, 1995).

60. Ewen Callaway, "The Discovery of Homo Floresiensis: The Tales of the Hobbit," *Nature* 514 (October 23, 2014): 422–26.

61. Stephen Jay Gould, *Rocks of Ages: Science and Religion in the Fullness of Life* (New York: Ballantine Books, 2002), 4.

62. Jerry Coyne, *Faith vs. Fact: Why Science and Religion Are Incompatible* (New York: Viking, 2015), 138.

63. Ibid., 99–100.

64. Ibid., 142–43.

65. Ibid., 137.

66. See David Hume's delightful and persuasive discussion of miracles in his *An Inquiry concerning Human Understanding* (New York: Bobbs-Merrill, 1955), 117–41; quote from 118.

67. Ibid., 192; emphasis in the original.

68. Darwin, *Origin of Species*, 62. The "face of nature" is a recurring metaphor in the *Origin* (62, 67, 73).

69. See Mary Hesse, *Models and Analogies in Science*, new ed. (Notre Dame, IN: Notre Dame University Press, 1970).

70. Bertrand Russell, "Knowledge by Acquaintance and Knowledge by Description," in his *Mysticism and Logic* (London: Allen & Unwin, [1917] 1963), 152–67.

71. Coyne, *Faith vs. Fact*, 229–39.

72. Friedrich Daniel Ernst Schleiermacher, *Über die Religion: Reden an die Gebildeten unter ihren Verächtern*, in *Kritische Gesamtausgabe*, ed. Hans-Joachim Birkner et al., 11 vols. to date (Berlin: Walter de Gruyter, 1980), 1:2.

73. J. B. S. Haldane, *Possible Worlds and Other Essays* (London: Chatto and Windus, 1927), 286.

BIBLIOGRAPHY

Abbott, Patrick, Jun Abe, John Alcock, Samuel Alizon, Joao A. C. Alpedrinha, Malte Andersson, Jean-Baptiste Andre, et al. "Inclusive Fitness and Eusociality." *Nature* 471 (2011): E1–E4.

Allen, Elizabeth, Barbara Beckwith, Jon Beckwith, Steven Chorover, David Culver, Margaret Duncan, Steven Gould, et al. "Letter to the Editor." *New York Review of Books* 22, no. 18 (1975): 43–44.

Anonymous. "Darwin on the Descent of Man." *Edinburgh Review* 134 (1871): 195–235.

Anonymous. "Mr. Darwin on the Descent of Man." *Times* (London), April 7, 1871.

Appel, Toby A. *The Cuvier-Geoffroy Debate: French Biology in the Decades before Darwin.* Oxford: Oxford University Press, 1987.

Artigas, M., Thomas F. Glick, and R. A. Martinez. *Negotiating Darwin: The Vatican Confronts Evolution, 1877–1902.* Baltimore, MD: Johns Hopkins University Press, 2006.

Babbage, Charles. *The Ninth Bridgewater Treatise: A Fragment.* London: John Murray, 1838.

Bölsche, Wilhelm. *Ernst Haeckel: Ein Lebensbild.* Berlin: Georg Bondi, 1909.

Borrello, Mark. *Evolutionary Restraints: The Contentious History of Group Selection.* Chicago: University of Chicago Press, 2010.

Bowler, Peter. *Evolution, the History of an Idea.* 3rd ed. Berkeley: University of California Press, 2003.

Bowler, Peter. *The Non-Darwinian Revolution: Reinterpreting a Historical Myth.* Baltimore, MD: Johns Hopkins University Press, 1988.

Bowler, Peter. *Theories of Human Evolution.* Baltimore, MD: Johns Hopkins University Press, 1986.

Boyle, Robert. *A Free Enquiry into the Vulgarly Received Notion of Nature.* Edited by Edward B. Davis and Michael Hunter. Cambridge: Cambridge University Press, 1996.

Brewster, David. "Review of Comte's 'Cours de Philosophie Positive.'" *Edinburgh Review* 67 (1838): 271–308.

Broad, C.D. *The Mind and Its Place in Nature*. London: Kegan Paul, 1925.

Brooke, John Hedley. "'Laws Impressed on Matter by the Creator?' The Origin and the Question of Religion." In *The Cambridge Companion to the "Origin of Species,"* edited by Michael Ruse and Robert J. Richards, 256–74. Cambridge: Cambridge University Press, 2009.

Browne, Janet. *Charles Darwin: The Power of Place*. New York: Knopf, 2002.

Browne, Janet. *Charles Darwin: Voyaging*. New York: Knopf, 1995.

Burchfield, Joseph D. *Lord Kelvin and the Age of the Earth*. New York: Science History Publications, 1975.

Burkhardt, Frederick, Sydney Smith, James Secord, and E. Janet Browne, eds. *The Correspondence of Charles Darwin*. Cambridge: Cambridge University Press, 1985–.

Callaway, Ewen. "The Discovery of Homo floresiensis: The Tales of the Hobbit." *Nature* 514 (October 23, 2014): 422–26.

Campbell, Donald. "'Downward Causation' in Hierarchically Organized Biological Systems." In *Studies in the Philosophy of Biology*, edited by Francisco Ayala and Theodosius Dobzhansky, 179–86. London: Macmillan, 1974.

Cannon, Walter F. "The Impact of Uniformitarianism: Two Letters from John Herschel to Charles Lyell, 1836–1837." *Proceedings of the American Philosophical Society* 105 (1961): 301–14.

Carlson, Elof. *Mendel's Legacy: The Origin of Classical Genetics*. Cold Spring Harbor, NY: Cold Spring Harbor Laboratory Press, 2004.

Carr, Edward Hallett. *What Is History?* New York: Vintage, 1961.

Carroll, Robert L. *Vertebrate Paleontology and Evolution*. New York: Freeman and Co., 1988.

Carroll, Sean B. *Endless Forms Most Beautiful: The New Science of Evo Devo*. New York: Norton, 2005.

Carus, Carl Gustav. *Lebenserinnerungen und Denkwürdigkeiten*. 4 vols. Leipzig: F. A. Brockhaus, 1865–66.

Carus, Carl Gustav. *Von den Ur-Theilen des Knochen- und Schalengerüstes*. Leipzig: Gerhard Fleischer, 1828.

Chambers, Robert. *Vestiges of the Natural History of Creation*. 3rd ed. London: Churchill, [1844] 1846.

Chancellor, John. "Humboldt's Personal Narrative and Its Influence on Darwin." http://darwin-online.org.uk/EditorialIntroductions/Chancellor_Humboldt.html.

Cleto, Fabio. "The Biological Drama: Darwinian Ethics in George Gissing's Fiction." *Gissing Journal* 28, no. 3 (1992): 1–13.

Cobbe, Frances Power. "Darwinism in Morals." *Theological Review* 8 (1871): 167–92.

Coleridge, Samuel Taylor. *Biographia Literaria.* Vol. 7 of *The Collected Works of Samuel Taylor Coleridge*, edited by James Engell and W. Jackson Bate. Princeton, NJ: Princeton University Press, 1983.

Colley, Linda. *Britons: Forging the Nation, 1707–1837.* New Haven, CT: Yale University Press, 1992.

Cornell, John. "God's Magnificent Law: The Bad Influence of Theistic Metaphysics on Darwin's Estimation of Natural Selection." *Journal of the History of Biology* 20 (1987): 381–412.

Cornell, John. "Newton of the Grassblade? Darwin and the Problem of Organic Teleology." *Isis* 77 (1986): 404–21.

Coyne, Jerry. *Faith vs. Fact: Why Science and Religion Are Incompatible.* New York: Viking, 2015.

Cuvier, Georges. *Essay on the Theory of the Earth.* Translated by Robert Kerr. Edinburgh: W. Blackwood, 1813.

Darwin, Charles. *The Autobiography of Charles Darwin, 1809–1882: With the Original Omissions Restored; Edited and with Appendix and Notes by his Grand-Daughter Nora Barlow.* London: Collins, 1958.

Darwin, Charles. *Beagle Diary.* Edited by R. D. Keynes. Cambridge: Cambridge University Press, 1988.

Darwin, Charles. *Charles Darwin's Natural Selection, Being the Second Part of His Big Species Book Written from 1856 to 1858.* Edited by R. C. Stauffer. Cambridge: Cambridge University Press, 1975.

Darwin, Charles. *Charles Darwin's Notebooks, 1836–1844.* Edited by Paul H. Barrett, Peter J. Gautrey, Sandra Herbert, David Kohn, and Sydney Smith. Ithaca, NY: Cornell University Press, 1987.

Darwin, Charles. *Darwin's Ornithological Notes.* Edited by Nora Barlow. *Bulletin of the British Museum (Natural History)*, Historical ser., 2, no. 7 (1963): 262.

Darwin, Charles. *The Descent of Man and Selection in Relation to Sex.* 2 vols. London: John Murray, 1871.

Darwin, Charles. *The Descent of Man and Selection in Relation to Sex.* 2nd ed. 2 vols. London: John Murray, [1871] 1874.

Darwin, Charles. *The Descent of Man and Selection in Relation to Sex.* Edited by James Moore and Adrian Desmond. 2nd ed. London: Penguin, [1879] 2004.

Darwin, Charles. *The Different Forms of Flowers on Plants of the Same Species.* London: John Murray, 1877.

Darwin, Charles. *The Effects of Cross and Self Fertilisation in the Vegetable Kingdom.* 2nd ed. London: John Murray, [1876] 1878.

Darwin, Charles. *The Expression of Emotions in Man and in Animals.* London: John Murray, 1872.

Darwin, Charles. *The Formation of Vegetable Mould, through the Action of Worms.* London: John Murray, 1881.

Darwin, Charles. *Geological Observations on South America. Being the Third Part of the Geology of the Voyage of the Beagle*. London: Smith Elder and Co., 1846.

Darwin, Charles. *Geological Observations on the Volcanic Islands Visited during the Voyage of H.M.S. Beagle, Together with Some Brief Notices of the Geology of Australia and the Cape of Good Hope. Being the Second Part of the Geology of the Voyage of the Beagle*. London: Smith Elder and Co., 1844.

Darwin, Charles. *Insectivorous Plants*. 2nd ed. London: John Murray, [1875] 1888.

Darwin, Charles. *Journal of Researches into the Geology and Natural History of the Various Countries Visited by H.M.S. Beagle*. London: Henry Coburn, 1839.

Darwin, Charles. *Journal of Researches into the Natural History and Geology of the Countries Visited during the Voyage of H.M.S. Beagle Round the World under the Command of Capt. Fitz Roy, R.N.* 2nd ed. London: John Murray, 1845.

Darwin, Charles. *A Monograph of the Fossil Balanidae and Verrucidae of Great Britain*. London: Printed for the Palæontographical Society, 1854.

Darwin, Charles. *A Monograph of the Fossil Lepadidae; or, Pedunculated Cirripedes of Great Britain*. London: Printed for the Paleontographical Society, 1851.

Darwin, Charles. *A Monograph of the Sub-Class Cirripedia, with Figures of All the Species. The Balanidge (or Sessile Cirripedes); the Verrucidae, and C.* London: Ray Society, 1854.

Darwin, Charles. *A Monograph of the Sub-Class Cirripedia, with Figures of All the Species. The Lepadidae; or Pedunculated Cirripedes*. London: Ray Society, 1851.

Darwin, Charles. "On the Movements and Habits of Climbing Plants." *Journal of the Linnaean Society of London (Botany)* 9 (1865): 1–118.

Darwin, Charles. *On the Origin of Species by Means of Natural Selection; or, The Preservation of Favoured Races in the Struggle for Life*. London: John Murray, 1859.

Darwin, Charles. *On the Origin of Species by Means of Natural Selection; or, The Preservation of Favoured Races in the Struggle for Life*. 3rd ed. London: John Murray, [1859] 1861; cf. 6th ed. (1872).

Darwin, Charles. *On the Various Contrivances by Which British and Foreign Orchids Are Fertilised by Insects*. London: John Murray, 1862.

Darwin, Charles. *The Origin of Species by Charles Darwin: A Variorum Text*. Edited by Morse Peckham. Philadelphia: University of Pennsylvania Press, 1959.

Darwin, Charles. "Pangenesis." *Nature* 3 (April 27, 1871): 502–3.

Darwin, Charles. *The Power of Movement in Plants*. London: John Murray, 1880.

Darwin, Charles. *The Structure and Distribution of Coral Reefs. Being the First Part of the Geology of the Voyage of the Beagle*. London: Smith Elder and Co., 1842.

Darwin, Charles. *The Variation of Animals and Plants under Domestication*. 2 vols. London: John Murray, 1868.

Darwin, Charles, ed. *The Zoology of the Voyage of H.M.S. Beagle, under the Command of Captain Fitz Roy, during the Years 1832 to 1836*. London: Smith Elder and Co., 1838-43.

Darwin, Charles, and Alfred Russel Wallace. *Evolution by Natural Selection*. Edited by Gavin de Beer. Cambridge: Cambridge University Press, 1958.

Darwin, Erasmus. *The Botanic Garden: A Poem in Two Parts*. London: Jones and Co., 1825.

Darwin, Erasmus. *The Economy of Vegetation*. London: J. Johnson, 1791.

Darwin, Erasmus. *The Temple of Nature*. London: J. Johnson, 1803.

Darwin, Erasmus. *Zoonomia; Or, The Laws of Organic Life*. 2 vols. London: J. Johnson, 1794-96.

Darwin, Francis, ed. *Foundations of the Origin of Species: Two Essays Written in 1842 and 1844*. Cambridge: Cambridge University Press, 1909.

Darwin, Francis. *Life and Letters of Charles Darwin*. 2 vols. London: John Murray, 1887.

Darwin, Francis, and A. C. Steward. *More Letters of Charles Darwin*. 2 vols. London: John Murray, 1903.

Dawkins, Richard. *A River Out of Eden*. New York: Basic Books, 1995.

Dawkins, Richard. *The Selfish Gene*. Oxford: Oxford University Press, 1976.

Day, Michael H. *Guide to Fossil Man*. 4th ed. Chicago: University of Chicago Press, 1986.

de Buffrénil, V., J. O. Farlow, and A. de Ricqlès, "Growth and Function of Stegosaurus Plates: Evidence from Bone Histology." *Paleobiology* 12 (1986): 459-73.

Dennett, Daniel. *Consciousness Explained*. New York: Little Brown, 1991.

Dennett, Daniel. *Darwin's Dangerous Idea*. New York: Simon and Schuster, 1995.

Desmond, Adrian, and James Moore. *Darwin's Sacred Cause: How a Hatred of Slavery Shaped Darwin's Views on Human Evolution*. New York: Houghton, Mifflin, Harcourt, 2009.

Desmond, Adrian. "Robert E. Grant: The Social Predicament of a Pre-Darwinian Transmutationist." *Journal of the History of Biology* 17 (1984): 189-223.

de Waal, Frans. *The Age of Empathy: Nature's Lessons for a Kinder Society*. New York: Crown, 2009.

Di Gregorio, Mario, and N. W. Gill, eds. *Charles Darwin's Marginalia*. New York: Garland, 1990.

Dijksterhuis, Eduard J. *The Mechanization of the World Picture*. Oxford: Oxford University Press, 1961.

Disraeli, Benjamin. *Lothair*. 2 vols. Leipzig: Bernhard Tauchnitz, 1870.

Dixon, Mike, and Gregory Radick. *Darwin in Ilkley*. Ikley: History Press, 2009.

Dobzhansky, Theodosius. "Nothing in Biology Makes Sense Except in the Light of Evolution." *American Biology Teacher* 35 (1973): 125–29.

Eckermann, Johann Peter. *Gespräche mit Goethe in den letzten Jahren seines Lebends*. 3rd ed. Berlin: Aufau-Verlag, 1987.

Ecklund, Elaine Howard. *Science vs. Religion: What Scientists Really Think*. Oxford: Oxford University Press, 2010.

Evans, Eric J. *The Forging of the Modern State: Early Industrial Britain, 1783–1870*. 3rd ed. Harlow, Essex: Longman, 2001.

Forster, Georg. *A Voyage Round the World in His Britannic Majesty's Sloop, Resolution, Commanded by Capt. James Cook, during the Years 1772, 3, 4, and 5*. 2 vols. London: B. White, 1777.

Goethe, Johann Wolfgang von. *Versuch die Metamorphose der Pflanzen zu erklären*. Gotha: Carl Wilhelm Ettinger, 1790.

Gould, Stephen Jay. "Eternal Metaphors of Palaeontology." In *Patterns of Evolution as Illustrated in the Fossil Record*, edited by A. Hallam, 13. New York: Elsevier, 1977.

Gould, Stephen Jay. *Rocks of Ages: Science and Religion in the Fullness of Life*. New York: Ballantine Books, 2002.

Gould, Stephen Jay, and Richard Lewontin. "The Spandrels of San Marco and the Panglossian Paradigm: A Critique of the Adaptationist Programme." *Proceedings of the Royal Society of London, Series B: Biological Sciences* 205 (1979): 581–98.

Gould, Stephen Jay. *The Structure of Evolutionary Theory*. Cambridge, MA: Harvard University Press, 2002.

Gray, Asa. *Darwiniana*. New York: Appleton, 1876.

Green, Richard, Johannes Krause, Adrian W. Briggs, Tomislav Maricic, Udo Stenzel, Martin Kircher, Nick Patterson, et al. "A Draft Sequence of the Neandertal Genome." *Science* 328 (May 7, 2010): 710–20.

Haldane, J. B. S. *Possible Worlds and Other Essays*. London: Chatto and Windus, 1927.

Hall, A. Rupert. *The Revolution in Science, 1500–1750*. London: Longman, 1983.

Hamilton, William. "The Genetical Evolution of Social Behaviour." *Journal of Theoretical Biology* 7 (1964): 1–52.

Hauser, Marc, Liane Young, and Fiery Cushman. "Reviving Rawls' Linguistic Analogy: Operative Principles and the Causal Structure of Moral Actions."

Vol. 2 of *Moral Psychology*, edited by W. Sinnott-Armstrong. Cambridge: MIT Press, 2007.

Hegel, George F. W. *Philosophy of Nature*. Oxford: Oxford University Press, [1817] 1970.

Henry, W. *The Elements of Experimental Chemistry*. 8th ed. London: Baldwin, Cradock and Joy, 1818.

Henslow, John S. *Descriptive and Physiological Botany*. London: Longman, Rees, Orme, Brown, Green, and Longman; and John Taylor, 1836.

Herbert, Sandra. *Charles Darwin, Geologist*. Ithaca, NY: Cornell University Press, 2005.

Herschel, John F. W. "Light." In *Encyclopedia Metropolitana*, edited by Edward Smedley, Hugh James Rose, and Henry John Rose. London: J. Griffin, 1827.

Herschel, John F. W. *Preliminary Discourse on the Study of Natural Philosophy*. London: Longman, Rees, Orme, Brown, Green, and Longman, [1830] 1831.

Herschel, John F. W. "Review of Whewell's History and Philosophy." *Quarterly Review* 135 (1841): 177–238.

Hesse, Mary. *Models and Analogies in Science*. New ed. Notre Dame, IN: Notre Dame University Press, 1970.

Hilton, Boyd. *A Mad, Bad and Dangerous People? England 1783–1846*. The New Oxford History of England. Oxford: Oxford University Press, 2006.

Hodge, Charles. *What Is Darwinism?* New York: Scribner's, 1874.

Hodge, M. Jonathan S. "Capitalist Contexts for Darwinian Theory: Land, Finance, Industry and Empire." *Journal of the History of Biology* 42 (2009): 399–416.

Holmes, John. "Literature and Science vs. History of Science." *Journal of Literature and Science* 5, no. 2 (2012): 67–71.

Humboldt, Alexander von. *Cosmos*. Translated by E. Sabine. 2 vols. London: Longman, Brown, Green & Longmans, 1846.

Humboldt, Alexander von. *Cosmos: A Sketch of the Physical Description of the Universe*. Translated by E. C. Otté. 5 vols. New York: Harper, n.d.

Humboldt, Alexander von. *Essai Géognostique sur le Gisement des Roches dans les Deux Hemispheres*. 2nd ed. Paris: F. G. Levrault, 1826.

Humboldt, Alexander von. *Fragmens de Géologie et de Climatologie Asiatiques*. 2 vols. Paris: A. Pihan DelaForest, 1831.

Humboldt, Alexander von. *Humboldt über das Universum: Alexander von Humboldt, Die Kosmosvorträge 1827/28*. Frankfurt a.M.: Insel Verlag, 1993.

Humboldt, Alexander von. *Political Essay on the Kingdom of New Spain*. Translated by John Black. 2 vols. London: Longman, Hurst, Rees, Orme, and Brown, 1911–22.

Humboldt, Alexander von, and Aimé Bonpland. *Personal Narrative of Travels to the Equinoctial Regions of the New Continent, during the Years 1799–1804*.

Translated by Helen Maria Williams. 7 vols. London: Longman, Hurst, Rees, Orme, and Brown, 1814–29.

Hume, David. *An Inquiry Concerning Human Understanding*. New York: Bobbs-Merrill, 1955.

Hume, David. *A Treatise of Human Nature*. Edited by D. J. Norton and M. J. Norton. Oxford: Oxford University Press, 2000.

Hutton, James. *Theory of the Earth, with Proofs and Illustrations*. Edinburgh: William Creech, 1795.

Huxley, Julian. *Evolution: The Modern Synthesis*. London: Allen and Unwin, 1942.

Huxley, Thomas Henry. "Darwin on the Origin of Species." *Westminster Review*, n.s., 17 (1860): 541–70.

Huxley, Thomas Henry. *Evidence as to Man's Place in Nature*. London: Williams and Norgate, 1863.

Huxley, Thomas Henry. "Evolution and Ethics." Vol. 9 of *Collected Essays*. New York: D. Appleton, 1902.

Huxley, Thomas Henry. "The Genealogy of Animals." Vol. 2 of *Collected Essays* by Thomas Henry Huxley. London: McMillan & Co., 1893.

Huxley, Thomas Henry. "On the Hypothesis That Animals Are Automata and Its History." *Fortnightly Review* 22 (1874): 555–89.

Huxley, Thomas Henry. "On the Morphology of the Cephalus Mullusca, as Illustrated by the Anatomy of Certain Heteropoda and Pteropoda Collected during the Voyage of the HMS Rattlesnake in 1846–50." *Philosophical Transactions of the Royal Society* 143, no. 1 (1853): 29–66.

Huxley, Thomas Henry. "On the Theory of the Vertebrate Skull: Croonian Lecture Delivered before the Royal Society, June 17, 1858." *Proceedings of the Royal Society* 9 (1857–59): 381–457.

James, William. "Does Consciousness Exist?" In *The Works of William James: Essays in Radical Empiricism*, edited by Frederick Burkhardt. Cambridge, MA: Harvard University Press, 1976.

James, William. *The Principles of Psychology*. 2 vols. New York: Henry Holt, 1890.

Jenkin, Fleeming. "The Origin of Species." *North British Review* 46 (1867): 277–318.

Johanson, Donald, and Maitland Edey. *Lucy: The Beginnings of Humankind*. New York: Simon & Schuster, 1990.

Kant, Immanuel, *Metaphysics of Ethics*. Translated by J. W. Semple. Edinburgh: T. Clark, 1836.

Kirby, William, and William Spence. *An Introduction to Entomology; or, Elements of the Natural History of Insects*. 2nd ed. 4 vols. London: Longman, Hurst, Reece, Orme, and Brown, 1815–28.

Laland, Kevin, Tobias Uller, Marc Feldman, Kim Sterelny, Gerd B. Müller, Armin Moczek, Eva Jablonka, et al. "Does Evolutionary Theory Need a Rethink?" *Nature* 514 (2013): 162–64.

Larson, Edward, and Larry Witham. "Leading Scientists Still Reject God." *Nature* 394 (July 23, 1998): 313.

Larson, Edward, and Larry Witham, "Scientists Are Still Keeping the Faith." *Nature* 386 (April 3, 1997): 435–36.

Laudan, Rachel. *From Mineralogy to Geology: The Foundations of a Science, 1650–1830.* Chicago: University of Chicago Press, 1987.

Leakey, Richard, and Roger Lewin. *Origins: The Emergence and Evolution of Our Species and Its Possible Future.* New York: Penguin, 1991.

Leakey, Richard, and Roger Lewin. *Origins Reconsidered: In Search of What Makes Us Human.* New York: Anchor Books, 1993.

Leuba, James. *The Belief in God and Immortality: A Psychological, Anthropological and Statistical Study.* Boston: Sherman, French, & Co., 1916.

Lewontin, R. C., Steven Rose, and Leon Kamin. *Not in Our Genes.* New York: Pantheon, 1984.

Litchfield, Henrietta. *Emma Darwin: A Century of Family Letters.* Cambridge: Privately Printed, 1905.

Lucas, John R. "Wilberforce and Huxley: A Legendary Encounter." *Historical Journal* 22 (1979): 313–30.

Lurie, E. *Louis Agassiz: A Life in Science.* Chicago: University of Chicago Press, 1960.

Lyell, Charles. *Principles of Geology: Being an Attempt to Explain the Former Changes in the Earth's Surface by Reference to Causes Now in Operation.* 3 vols. London: John Murray, 1830–33.

Mackintosh, James. *Dissertation on the Progress of Ethical Philosophy.* Edinburgh: Adam and Charles Black, 1836.

Malthus, Thomas Robert. *An Essay on the Principle of Population.* London: Printed for J. Johnson in St. Paul's Church-Yard, 1798. Reprint, New York: Macmillan, 1996.

Malthus, Thomas Robert. *Essay on a Principle of Population.* 6th ed. London: John Murray, 1826.

Martineau, Harriet. *Chamber's Encyclopedia of English Literature.* Edited by David Patrick. 2nd ed. 3 vols. London: W. R. Chambers, 1903.

Mayr, Otto. *Authority, Liberty, and Automatic Machinery in Early Modern Europe.* Baltimore, MD: Johns Hopkins University Press, 1989.

McGinn, Colin. *The Mysterious Flame: Conscious Minds in a Material World.* New York: Basic Books, 2000.

M'Culloch, J. R. *The Principles of Political Economy.* London: Alex, Murray and Son, 1825.

Mill, John Stuart, "Review of the Works of Samuel Taylor Coleridge." *London and Westminster Review* 33 (1840): 257–302.

Milne-Edwards, Henri. *Introduction à la Zoologie Générale*. Paris: Victor Masson, 1851.

Mivart, St. George Jackson. "Review of the Descent of Man." *Quarterly Review* 131 (1871): 47–90.

Morrell, Jack, and Arnold Thackray. *Gentlemen of Science: Early Years of the British Association for the Advancement of Science*. Oxford: Oxford University Press, 1981.

Myers, Frederick W. H. "George Eliot." *Century Magazine* (November 1881): 57–63.

Nagel, Thomas. *Mind and Cosmos: Why the Materialist Neo-Darwinian Conception of Nature Is Almost Certainly False*. Oxford: Oxford University Press, 2012.

Nicholson, Malcolm. "Historical Introduction" to *Personal Narrative of a Journey to the Equinoctial Regions of the New Continent*, by Alexander von Humboldt. Translated by Jason Wilson. London: Penguin, 1995.

Nordenskiöld, Erik. *History of Biology*. New York: Knopf, [1920–24] 1935.

Nowak, Martin A., Corina E. Tarnita, and Edward O. Wilson. "The Evolution of Eusociality." *Nature* 466 (2010): 1057–62.

Ospovat, Dov. *The Development of Darwin's Theory*. Cambridge: Cambridge University Press, 1981.

Owen, Richard. *On the Archetype and Homologies of the Vertebrate Skeleton*. London: John Van Voorst, 1848.

Owen, Richard. *On the Nature of Limbs*. London: John Van Voorst, 1849.

Paley, William. *Moral and Political Philosophy*. In *The Works of William Paley*. New ed. 7 vols. London: C. and J. Rivington, 1825.

Paley, William. *Natural Theology; or, Evidences of the Existence and Attributes of the Deity*. 12th ed. London: J. Faulder, [1802] 1809.

Paley, William. *View of the Evidences of Christianity*. London: Faulder, 1794.

Parkes, Samuel. *The Chemical Catechism, with Notes, Illustrations and Experiments*. London: Baldwin, Cradock and Joy, 1818.

Pennisi, Elizabeth. "Genomics: Encode Project Writes Eulogy for Junk DNA." *Science* 337 (September 7, 2012): 1159–61.

Popper, Karl, and John Eccles. *The Self and Its Brain*. New York: Springer-Verlag, 1977.

Powell, Baden. *Essays on the Spirit of the Inductive Philosophy*. London: Longman, Brown, Green, and Longmans, 1855.

Prodger, Phillip. *Darwin's Camera: Art and Photography in the Theory of Evolution*. Oxford: Oxford University Press, 2009.

Provine, William. *The Origins of Theoretical Population Genetics*. Chicago: University of Chicago Press, 1971.

Richards, Robert J. *Darwin and the Emergence of Evolutionary Theory of Mind and Behavior*. Chicago: University of Chicago Press, [1987] 1988.

Richards, Robert J. *The Meaning of Evolution: The Morphological Construction and Ideological Reconstruction of Darwin's Theory*. Chicago: University of Chicago Press, 1992.

Richards, Robert J. *The Romantic Conception of Life: Science and Philosophy in the Age of Goethe*. Chicago: University of Chicago Press, [2002] 2003.

Richards, Robert J. *The Tragic Sense of Life: Ernst Haeckel and the Struggle over Evolutionary Thought*. Chicago: University of Chicago Press, 2008.

Richards, Robert J. *Was Hitler a Darwinian? Disputed Questions in the History of Evolutionary Theory*. Chicago: University of Chicago Press, 2013.

Richards, Robert J., and Richard Lewontin. "Darwin and Progress." *New York Review of Books* 52, no. 20 (December 15, 2005), letters.

Robinson, Henry Crabb. *Diary, Reminiscences, and Correspondence*. Edited by Thomas Sadler. 3rd ed. 2 vols. London: Macmillan and Co., 1872.

Ross, Sydney. "Scientist: The Story of a Word." *Annals of Science* 18, no. 2 (1964): 65–85.

Rudwick, Martin J. S. *Bursting the Limits of Time*. Chicago: University of Chicago Press, 2005.

Rudwick, Martin J. S. *The Meaning of Fossils*. New York: Science History Publications, 1972.

Rudwick, Martin J. S. "The Strategy of Lyell's *Principles of Geology*." *Isis* 61 (1969): 5–33.

Rupke, Nicolass. *Richard Owen, Biology without Darwin*. 2nd ed. Chicago: University of Chicago Press, 2009.

Ruse, Michael. "Adaptive Landscapes and Dynamic Equilibrium: The Spencerian Contribution to Twentieth-Century American Evolutionary Biology." In *Darwinian Heresies*, edited Abigail Lustig, Robert J. Richards, and Michael Ruse, 131–50. Cambridge: Cambridge University Press, 2004.

Ruse, Michael. *The Cambridge Encyclopedia of Darwin and Evolutionary Thought*. Cambridge: Cambridge University Press, 2013.

Ruse, Michael. "Charles Darwin and Artificial Selection." *Journal of the History of Ideas* 36 (1975): 339–50.

Ruse, Michael. "Charles Darwin and Group Selection." *Annals of Science* 37 (1980): 615–30.

Ruse, Michael. "Charles Robert Darwin and Alfred Russel Wallace: Their Dispute over the Units of Selection." *Theory in Biosciences* 132 (2013): 215–24.

Ruse, Michael. *Darwin and Design: Does Evolution Have a Purpose?* Cambridge, MA: Harvard University Press, 2003.

Ruse, Michael. "Darwin and Mechanism: Metaphor in Science." *Studies in History and Philosophy of Biology and Biomedical Sciences* 36 (2005): 285–302.

Ruse, Michael. "The Darwinian Revolution: Rethinking Its Meaning and Significance." *Proceedings of the National Academy of Sciences* 106 (2009): 10040–7.

Ruse, Michael. *The Darwinian Revolution: Science Red in Tooth and Claw*. Chicago: University of Chicago Press, 1979.

Ruse, Michael. *Darwinism as Religion: What Literature Tells Us about Evolution*. New York: Oxford University Press, 2016.

Ruse, Michael. "Darwin's Debt to Philosophy: An Examination of the Influence of the Philosophical Ideas of John F. W. Herschel and William Whewell on the Development of Charles Darwin's Theory of Evolution." *Studies in History and Philosophy of Science* 6 (1975): 159–81.

Ruse, Michael. *The Evolution-Creation Struggle*. Cambridge, MA: Harvard University Press, 2005.

Ruse, Michael. *The Evolution Wars*. 2nd ed. Millerton, NY: Greyhouse Publishing, 2009.

Ruse, Michael. *The Gaia Hypothesis: Science on a Pagan Planet*. Chicago: University of Chicago Press, 2013.

Ruse, Michael. "Kant and Evolution." In *Theories of Generation*, edited by J. Smith, 402–15. Cambridge: Cambridge University Press, 2006.

Ruse, Michael. *Monad to Man: The Concept of Progress in Evolutionary Biology*. Cambridge, MA: Harvard University Press, 1996.

Ruse, Michael. "Natural Selection in the Origin of Species." *Studies in History and Philosophy of Science* 1 (1971): 311–52.

Ruse, Michael, ed. *Philosophy after Darwin: Classic and Contemporary Themes*. Princeton, NJ: Princeton University Press, 2009.

Ruse, Michael. *The Philosophy of Human Evolution*. Cambridge: Cambridge University Press, 2012.

Ruse, Michael. "Population Genetics." In *The Cambridge Encyclopedia of Darwin and Evolutionary Thought*, edited by Michael Ruse, 273–81. Cambridge: Cambridge University Press, 2013.

Ruse, Michael. *Science and Spirituality: Making Room for Faith in the Age of Science*. Cambridge: Cambridge University Press, 2010.

Ruse, Michael. *Sociobiology: Sense or Nonsense?* Dordrecht: Reidel, 1979.

Ruse, Michael, and E. O. Wilson. "Moral Philosophy as Applied Science." *Philosophy* 61 (1986): 173–92.

Russell, Bertrand. *The Analysis of Mind*. London: George Allen & Unwin, [1921] 1978.

Russell, Bertrand. "Knowledge by Acquaintance and Knowledge by Description." In *Mysticism and Logic*, by Bertrand Russell. London: Allen & Unwin, [1917] 1963.

Russell, Edwin S. *Form and Function: A Contribution to the History of Animal Morphology*. London: John Murray, 1916.

Schleiermacher, Friedrich Daniel Ernst. *Über die Religion: Reden an die Gebildeten unter ihren Verächtern*. Vol. 1 of *Kritische Gesamtausgabe*, edited by Hans-Joachim Birkner et al. Berlin: Walter de Gruyter, 1980–.

Schweber, Sam. "Darwin and the Political Economists: Divergence of Character." *Journal of the History of Biology* 13 (1980): 195–289.

Schweber, Sam. "The Wider British Context of Darwin's Theorizing." In *The Darwinian Heritage*, edited by David Kohn, 35–69. Princeton, NJ: Princeton University Press, 1985.

Sebright, John. *The Art of Improving the Breeds of Domestic Animals in a Letter Addressed to the Right Hon. Sir Joseph Banks, K.B.* London: John Harding, 1809.

Secord, James. *Victorian Sensation: The Extraordinary Publication, Reception, and Secret Authorship of "Vestiges of the Natural History of Creation."* Chicago: University of Chicago Press, 2000.

Sedgwick, Adam. *Discourse on the Studies at the University of Cambridge*. 5th ed. Cambridge: Cambridge University Press, 1850.

Sedgwick, Adam. "Objections to Mr. Darwin's Theory of the Origin of Species." In *Darwin and His Critics*, edited by David Hull, 159–66. Cambridge, MA: Harvard University Press, 1973.

Sedgwick, Adam. "Presidential Address to the Geological Society." *Proceedings of the Geological Society of London* 1 (1831): 281–316.

Sedgwick, Adam. "Vestiges." *Edinburgh Review* 82 (1845): 1–85.

Sepkoski, David. *Rereading the Fossil Record: The Growth of Paleobiology as an Evolutionary Discipline*. Chicago: University of Chicago Press, 2012.

Sepkoski, David, and Michael Ruse, eds. *The Paleobiological Revolution: Essays on the Growth of Modern Paleontology*. Chicago: University of Chicago Press, 2009.

Shaw, George Bernard. Preface to *Back to Methuselah*. London: Penguin Books, [1921] 1961.

Sidgwick, Henry. "The Theory of Evolution in Its Application to Practice." *Mind* 1 (1876): 52–67.

Sloan, Phillip. "Darwin, Vital Matter, and the Transformation of Species." *Journal of the History of Biology* 19 (Fall 1986): 369–445.

Sloan, Phillip. "'The Sense of Sublimity': Darwin on Nature and Divinity." *Osiris*, 2nd ser., 16 (2001): 251–69.

Smith, Adam. *An Inquiry into the Nature and Causes of the Wealth of Nations*. London: W. Strahan and T. Cadell, 1776.

Snyder, Laura. *Reforming Philosophy*. Chicago: University of Chicago Press, 2006.

Sober, Elliott. *Did Darwin Write the "Origin" Backwards? Philosophical Essays on Darwin's Theory*. Amherst, NY: Prometheus Books, 2011.

Stoddart, David R., ed. "Coral Islands by Charles Darwin." *Atoll Research Bulletin* 88 (December 1962): 1–20.

Sulloway, Frank. "Darwin's Conversion: The Beagle Voyage and Its Aftermath." *Journal of the History of Biology*. 15 (1982): 325–96.

Tattersall, Ian. *The Fossil Trail: How We Know What We Think We Know about Human Evolution*. Oxford: Oxford University Press, 1995.

Thatcher, Margaret. "No Such Thing as Society: Interview with Margaret Thatcher." By Douglas Keay. *Woman's Own* (September 23, 1987).

Tooby, John, and Leda Cosmides. "The Psychological Foundations of Culture." In *The Adapted Mind*, edited by Jerome Barkow, Leda Cosmides, and John Tooby, 19–136. New York: Oxford University Press, 1992.

Trevor-Roper, Hugh. *History and the Enlightenment*. New Haven, CT: Yale University Press, 2010.

Trivers, Robert. "The Evolution of Reciprocal Altruism." *Quarterly Review of Biology* 46 (1971): 35–57.

Van Dyke, F., D. C. Mahan, J. K. Sheldon, and R. H. Brand. *Redeeming Creation: The Biblical Basis for Environmental Stewardship*. Downer's Grove, IL: InterVarsity Press, 1996.

Vesalius, Andreas. *De Corporis Humani Fabrica, Libri Septem*. Basel: Oporinus, 1543.

Wade, Michael J. "A Critical Review of the Models of Group Selection." *Quarterly Review of Biology* 53 (1978): 101–14.

Wallace, Alfred Russel. *Contributions to the Theory of Natural Selection: A Series of Essays*. London: Macmillan, 1870.

Wallace, Alfred Russel. *Darwinism: An Exposition of the Theory of Natural Selection, with Some of Its Applications*. Cambridge: Cambridge University Press, 1889.

Wallace, Alfred Russel. "The Origin of Human Races and the Antiquity of Man Deduced from the Theory of 'Natural Selection.'" *Journal of the Anthropological Society of London* 2 (1864): 157–87.

Wallace, Alfred Russel. "Sir Charles Lyell on Geological Climates and the Origin of Species." *Quarterly Review* 126 (1869): 359–94.

Walls, Laura Dassow. *The Passage to Cosmos: Alexander von Humboldt and the Shaping of America*. Chicago: University of Chicago Press, 2009.

Watson, J. D., and F. H. Crick. "Molecular Structure of Nucleic Acids: A Structure for Deoxyribose Nucleic Acid." *Nature* 171 (1953): 737–38.

Weikardt, Richard. "A Recently Discovered Darwin Letter on Social Darwinism." *Isis* 86 (1995): 609–11.

Weismann, August. *The Germ-Plasm: A Theory of Heredity.* Translated by W. N. Parker and H. Rönnfeldt. New York: Scribner's Son, 1893.

Weismann, August. "The Supposed Transmission of Mutilations." *Essays upon Heredity and Kindred Biological Problems,* by August Weismann, translated and edited by Edward B. Poulton, Selmar Schönland, and Arthur E. Shipley, 421–48. 2 vols. Oxford: Clarendon Press, 1889.

West, Stuart A., and Andy Gardner. "Adaptation and Inclusive Fitness." *Current Biology* 23 (2013): R577–R584. doi: 10.1016/j.cub/2013.05.031.

West, Stuart A., A. S. Griffin, and Andy Gardner. "Social Semantics: How Useful Has Group Selection Been?" *Journal of Evolutionary Biology* 21 (2007): 374–85. doi: 10.111/j/1420.9109.2007.01458.x.

Whewell, William. *Astronomy and General Physics Considered with Reference to Natural Theology (Bridgewater Treatise).* Philadelphia: Carey, Lea & Blanchard, 1833.

Whewell, William. *History of the Inductive Sciences, from Earliest to Present Times.* 3rd ed. 2 vols. London: Parker & Son, [1837] 1857.

Whewell, William. *Philosophy of the Inductive Sciences.* London: Parker, 1840.

Whewell, William. *The Plurality of Worlds.* London: Parker, 1853.

Whitehead, Albert North. *Process and Reality.* New York: Free Press, 1979.

Wilkinson, John. *Remarks on the Improvement of Cattle, etc. in a Letter to Sir John Sanders Sebright, Bart. M.P.* Nottingham, 1820.

Williams, George C. *Adaptation and Natural Selection.* Princeton, NJ: Princeton University Press, 1966.

Wilson, David Sloan, and E. O. Wilson. "Rethinking the Theoretical Foundation of Sociobiology." *Quarterly Review of Biology* 82 (2007): 327–48.

Wilson, Edward O. *Sociobiology: The New Synthesis.* Cambridge, MA: Harvard University Press, 1975.

Wilson, Leonard G. *Charles Lyell, the Years to 1841: The Revolution in Geology.* New Haven, CT: Yale University Press, 1972.

Wilson, Leonard G., ed. *Sir Charles Lyell's Scientific Journals on the Species Question.* New Haven, CT: Yale University Press, 1970.

Wordsworth, William. Preface to *Lyrical Ballads.* Quoted by Gillian Beer, "Darwin's Reading and the Fictions of Development," in *The Darwinian Heritage,* edited by David Kohn, 543–88. Princeton, NJ: Princeton University Press, 1985.

INDEX